ALFA ROMEO
SPORTING SALOONS

JOHN TIPLER

SUTTON PUBLISHING LIMITED

Sutton Publishing Limited
Phoenix Mill · Thrupp · Stroud
Gloucestershire · GL5 2BU

First published 1999

Copyright © John Tipler, 1999

Cover photographs:
Title page photograph: Ian Brookfield leads the pack around Snetterton's Russell bend.

British Library Cataloguing in Publication Data
A catalogue record for this book is available from the British Library.

ISBN 0-7509-2078-5

Typeset in 10/11 Bembo.
Typesetting and origination by Sutton Publishing Limited.
Printed in Great Britain by Ebenezer Baylis, Worcester.

Alfa Romeo 146 2.0 ti.

Contents

	Introduction	5
1.	History of Alfa Romeo	9
2.	Mass Production	17
3.	From Giulietta to Giulia	39
4.	Mass Appeal	65
5.	Famous Names Return	89
6.	The Modern World	117
	Acknowledgements	158
	Index	159

A 1995 'wide-body' Alfa 155 V6 pictured in Cumbria.

INTRODUCTION

Alfa Romeo has always been associated with sports cars and grand tourers. Most film-goers will be familiar with the 1967 movie *The Graduate*, in which Dustin Hoffman drives from adolescence into manhood in a Duetto Spider. Fans of touring car racing will recall the lightweight Giulia GTA coupés trading paint and door handles with BMWs and Lotus Cortinas during the decade that spanned the mid-'60s to the mid-'70s. Alfa Romeo won the world sports car title in 1975 and 1977 with its flat 12-engined Tipo 33 prototypes. All the while enthusiasts could lust after the elegant Pininfarina-designed Spider and Bertone-styled Sprint GT coupé.

But what of their saloon cars, the Berlinas, unsung heroes of the road, which raised the level of family motoring from the mundane to the positively enthralling? The Berlinas – the name means four-door saloon in Italian – were always Alfa's bread-and-butter cars, while the exotic Spiders and GTs provided the jam. The Berlinas have a loyal and devoted following among family drivers who need four doors and a decent-sized back seat. But they have tended to be overlooked in the British and North American markets by enthusiasts who see them as being somehow less charismatic than their two-door siblings; in contrast, they have been neglected by the car-buying public at large as being too exotic. Alfa's often quirky design foibles can be forgiven in an extrovert GT car, but are less tolerable in a family hack.

Yet the Berlinas have always been noted for their sporty handling and decent performance, because they have always used the same suspension and running gear as the GTs and sports cars. Put that the other way round: the sporting models have generally been based on the saloon platform. Indeed, the saloons often handle better because the saloon bodyshell is stiffer and less flexible than that of the coupé or Spider.

In most respects, Alfa saloons were just as charismatic as their siblings. They were made in greater volumes, of course, but the people who own them regard them just as highly as the high-profile models. Today the BMW 3-series saloons are widely regarded as the benchmark sports saloons, but back in the 1950s and '60s it was Alfa Romeo that set the pace. The Giulietta and Giulia Berlinas were compact, lightweight, and endowed with high-revving twin-cam engines, allied to a wonderfully slick gearchange which has never been bettered. A sporting suspension set-up gave the enthusiastic family motorist sports car handling and performance. At a price, of course, but then an Alfa was viewed in the same prestigious light as a Jaguar with which it compared in price. And outside Italy, of course, an Alfa was more unusual.

Alfa Romeo's heritage as a builder of world-class sports and touring machines dates back to 1910. The company emerged in 1950 on the lower rungs of the volume car manufacturing ladder. Its reputation was founded on race success, having dominated the Grand Prix scene in the early days of the Formula One World Championship between 1949 and 1952. Its first really notable saloon car was the Vila d'Este of 1948, which was a very grand vehicle indeed.

Chapter two covers the 1900 model, the company's first series production four-door saloon, launched in 1950. This was one of the first cars to be built as a unit-construction monocoque with no separate chassis. Next, in 1954, came the Giulietta Berlina, which set new standards for handling and performance with its lightweight, effective suspension and 1300cc all-alloy twin-cam engine. Alongside the factory-produced Berlina were marketed the Bertone-produced Giulietta Sprint and Pininfarina-built Spider. Fully-trimmed coupé and roadster bodies were made at the Bertone and Pininfarina factories and shipped to Alfa HQ for the drivetrains and running gear to be installed, while the saloon cars were all made in-house.

Alfa has also usually built bigger six-cylinder saloons alongside its compact models. Aimed at the 'luxury' market, these were never available in large quantities and were not particularly successful. Again, it is fashionable in the motoring press to deride barges like the Alfetta-based Alfa 6, but it can be said in their defence that in a market segment striving to conform with Daimler-Benz products, they did at least offer something out of the ordinary. They were broadly similar in appearance to whatever was the regular-sized family saloon in the range.

Chapter three takes us into the 1960s with the Giulia TI and Super saloons, on which were based the Sprint GT derivatives and the Duetto Spider. Production moved from the old Portello plant to a greenfield site at Arese on the outskirts of Milan.

These compact four-door saloons remained in production from 1962 to 1973, using 1300 and 1600cc engines. Because of its quirky yet efficient styling, the Giulia Super is growing in popularity today among the classic car fraternity. In 1968 Alfa facelifted the bodyshell and fitted bigger engines, and the cars were known simply as the 1750 and 2000 Berlina. They were less popular than the Giulia Super, and this model outlasted its larger siblings, remaining in production until 1978. I have included photographs of a number of Alfa Romeo's contemporary coupés by way of comparison.

Chapter four concentrates on Alfa Romeo's biggest hit of the 1970s, the little Alfasud. They constructed a new factory near Naples to build it, and as soon as it was released in 1972 the flat-four powered two-door Alfasud set new standards for front-wheel-drive handling in the small car market. Its successor, the 33, was far less charismatic, but remains Alfa's most numerous model to date. The 33 was also available as a Pininfarina-built station-wagon.

Chapter five picks up the story in 1972 with the introduction of the four-door Alfetta saloon. This model also received a warm welcome, along with its Giugiaro-designed

Surprisingly, the boxy shape of the Giulia Super was the product of wind-tunnel testing, and the steeply raked windscreen, cut-off Kamm tail and attractive flutings along the car's flanks and roofline promoted a drag coefficient figure of cd 0.33 – far superior to that of a host of supposedly aerodynamic coupés.

The model was introduced in 1962 as the Giulia TI, and its platform and running gear formed the basis for the subsequent Duetto Spider and Sprint GT models. The Spider was not replaced until 1993, so the concept of the Giulia's componentry endured for a long time by motor industry standards. The Giulia Super itself went out of production in 1978.

The Giulia TI was powered by the 1570cc Alfa Romeo twin-cam, and it was soon joined by a sporting version known as the TI/Super, which had perspex rear windows, twin Weber carburettors and alloy wheels. It was soon in action on circuits the world over, usually locked in combat with the ubiquitous Lotus Cortinas.

As these models become rarer they are sought by Alfisti and restored to pristine condition, and they have been known to take outright honours at National Alfa Day concours events.

GTV sister. There were 1600, 1800 and 2-litre engine options, and the gearbox was mounted in the back axle for better weight distribution. The model was named after the Type 158 Alfetta Grand Prix car of 1948–52 which had a similar transmission arrangement. In 1980 the new Giulietta was launched. This model shared the Alfetta's running gear and was really no more than the same car with a shorter boot. In the background was the Alfa 6, a big six-cylinder limo that won few plaudits. It was followed by the Alfa 90, which was basically an Alfetta with the 2.5-litre V6 motor installed.

The chapter also includes the 75 model, named – or numbered – as a tribute to Alfa Romeo's seventy-five years as a motor manufacturer. It was available with 1800cc,

2-litre TwinSpark, 2.5- and 3-litre V6 engines, as well as a diesel version. It was the last of the rear-wheel-drive Alfas. Chapter five also sees the launch of two small saloons in the same class as the old Alfasud and 33 but featuring radically new styling. These were the two-door 145 hatchback and four-door 146 saloon. The two smaller cars started off using the flat-four 'Sud-derived engine, transferring to the straight-four twin overhead cam motor in 1996. There were three engine sizes – 1.6, 1.8 and 2-litres. The 156 model, with its mixture of modern design and traditional styling cues, was launched in 1997 to rave reviews, and promptly won the European 'Car of the Year' award by a substantial margin. The motor industry pundits expect that the 156 will dominate the sports saloon market for some time.

Chapter six ushers in the 164 model that was a Pininfarina-styled saloon, somewhat larger than the 75, and its 2-litre TwinSpark or 3-litre V6 or diesel engines drove through the front wheels. In a rare flash of automotive synergy it shared the same floorpan as the Saab 9000, Lancia Thema and Fiat Croma. While the 164 was hailed as a huge success by motor industry pundits, the 75's replacement, the 155, received a cooler reception. Launched in 1993, with the same compact dimensions as the 75, the 155's 1.8-, 2- TwinSpark and 2.5-litre V6 engines also drove through the front wheels. The top-of-the-range 166 effectively replaced the 164 in 1998.

The Alfa 33 succeeded the Alfasud and was built at the same Pomigliano d'Arco factory near Naples.

CHAPTER ONE

HISTORY OF ALFA ROMEO

Alfa Romeo's saloon cars have always been charismatic and often idiosyncratic but the company's reputation is really founded on its great feats both on the world's race circuits and in gruelling endurance events, many of which pre-date the modern production era. Its philosophy has always been to build a range of sports and touring cars to be marketed alongside the saloons. This book represents a thorough appraisal of all the company's post-war saloon cars. But for the uninitiated, it's a good idea to set them in the context of Alfa Romeo's long and eventful history.

Although the Alfa lineage goes back almost as far as the motor car itself, it wasn't until the early 1950s that Alfa Romeo joined Lancia and Fiat as one of Italy's top three volume car producers. It was at this point that the firm moved into the modern world of production-line unit-construction with the 1900 Berlinas, which were followed soon afterwards by the bread-and-butter Giulietta models.

Alfa Romeo is actually Italy's third oldest sporting manufacturer, preceded only by Fiat and Itala. It was founded in 1910 as Anonima Lombarda Fabrica Automobili, which provided that immortal acronym, ALFA. The company was originally set up in Naples in 1906 as a subsidiary of the French Darracq concern, to sell off surplus cars, but the following year the operation transferred to the labour-rich Milanese suburb of Portello. Insufficient demand for these fragile, underpowered French cars broke the fledgling company, and it was reformed as ALFA by Ugo Stella, with Giuseppe Merosi as chief designer.

The first ALFAs were quite different from the pretty Darracqs, being large and robust, with powerful engines and decent brakes, and much more appropriate for the indifferent quality of the majority of Italian roads in those days. Merosi's first two cars were the 4.1-litre 24hp model and the 2.4-litre 15hp machine. Their engines were not so far removed from the layout that Alfa Romeo has remained faithful to by and large throughout its history, with twin overhead camshafts operating two rows of inclined valves, and hemispherical combustion chambers.

From the outset ALFA's emblem comprised the familiar red cross of St George, which was the arms of the city of Milan, combined with the medieval shield adornment of the Visconti family, which is a serpent devouring a child. The four-leaf clover or quadrifoglio motif that has always featured on Alfa Romeo competition cars appeared on engine and chassis plates.

ALFA was pretty quickly off the mark into the competition fray, entering two 24hp cars in the Targa Florio, the arduous Sicilian road-race, and although one car led the

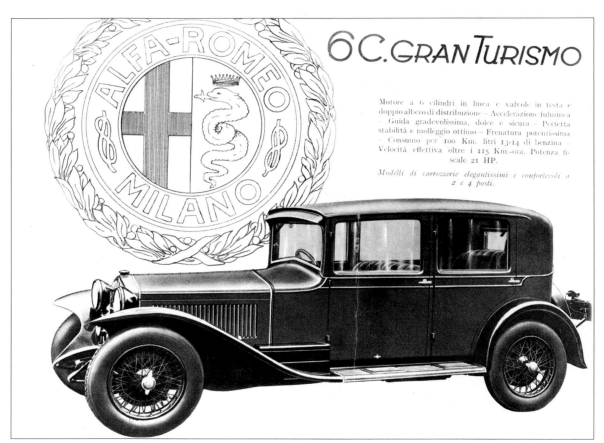

The straight-six Alfa Romeo 6C 1750 Gran Turismo was produced between 1929 and 1933. It was a handsome four-seater with four doors, capable of 70mph.

race, both eventually retired. ALFA's first taste of competition success came in 1913 when Campari and Franchini came third and fourth in the Coppa Florio. If the advent of the First World War prevented Merosi's promising four-cylinder twin-cam Grand Prix car being raced, it also had the beneficial effect of introducing prosperous Milanese mining engineer Nicola Romeo to ALFA. By arrangement, Ing. Romeo took over the ALFA plant in 1915 to produce compressors, tractors and Isotta aircraft engines for the war effort. When the war was over, Nicola Romeo became managing director, and the company was called Alfa Romeo thereafter.

Two pre-war cars took part in the 1919 Targa Florio without success, but things improved from the early 1920s when the works Alfa Romeo team was run by one of its drivers, Enzo Ferrari. Race victories began with Campari's wins at Mugello in 1920 and 1921, while Sivocci won the Targa Florio the following year driving Merosi's new 3-litre six-cylinder type-RL. In 1924 Vittorio Jano's first design, the 2-litre supercharged straight-eight P2, won first time out at Cremona with Antonio Ascari at the wheel, and its first Grand Prix at Lyons driven by Campari. Ascari was killed in his

This 1750 Gran Turismo four-door saloon was fitted with a larger body – note the rear three-quarter side window – and incorporated the spare wheel on the running board instead of at the rear of the car.

P2 the following year at Montlhéry, but the Alfa team had won sufficient races to gain them the first ever World Championship. This success was marked by the addition of a laurel wreath to the border of the Alfa Romeo badge.

Whereas Merosi's road cars of the early '20s were conceptually much the same as his pre-war cars – robust and heavy and hardly qualifying as sports cars, those produced in 1927 by his successor Jano were different altogether, and much more closely related to the racing cars. The engines were stripped of their superchargers and lost a couple of cylinders, and at first had only a set of vertical valves and a single camshaft. They were endowed with lightweight, open-top bodies mounted on race-developed chassis, and designated the 6C 1500 and 6C 1750, soon to be superseded by twin-cam Super Sport and 100mph Gransport versions with superchargers. Alfa Romeo's – and arguably the world's – first grand touring car appeared next as the saloon-bodied version known as the 1750 Gran Turismo.

Enzo Ferrari established the racing headquarters at Modena in 1929, from where he ran the works Alfas and maintained customers' competition cars. In the early 1930s Alfa Romeo virtually dominated international competition. The car that accomplished so much was Jano's 2.3-litre straight-eight 8C 2300, which won Le Mans in four consecutive years from 1931 to 1935. The Grand Prix version of the 8C won its first race, the 1931 European Grand Prix at Monza, and it was known as the Monza from then onwards. What was remarkable about the design of this engine was that it was basically two four-cylinder twin-cam engines facing each other with the camshaft drive

By 1934 the coachwork that clad the Gran Turismo chassis was more rounded and enveloped the car more completely. As was common practice at the time, the rear doors were hinged at the C-pillar. This model was fitted with the 6C 2300 engine, and its wheels covered in huge one-piece discs.

in the centre. This curious layout allowed for shorter cams and cranks, and was therefore, in theory, more reliable. This proved to be the case, for when fitted to the P3 Grand Prix car, it clocked up at least forty major successes between 1932 and 1935, most notable of which was probably Nuvolari's victory at the Nürburgring in 1935 against the mighty German 'silver arrows'.

By way of counter-attacking the Mercedes and Auto Unions, Scuderia Ferrari introduced the fearsome twin-engined Bimotore in 1935. It used two P3 engines located behind and in front of the driver, who sat on top of a three-speed gearbox that drove the rear wheels. The car was capable of 200mph in a straight line, but a ravenous appetite for tyres marked it as a failure.

Meanwhile things were not going well for the company financially. In 1933 ownership passed from the Banco di Sconto into government receivership and Alfa Romeo was refloated with its sights set on diverse commercial markets, including trucks, coaches, marine and aircraft engines. Cars for the domestic market took the form of handsome but hardly outstanding four-door saloons, powered by a straight-six engine of 1900cc or 2300cc. The 6C 2300's box-section chassis was considerably lighter than the C-section girders of its 8C 2300 predecessor, and its performance was sufficient to sustain a market which could keep the factory going as a commercial proposition. Short-wheelbase sports chassis were offered with bodies by Zagato, Castagna, Farina and Touring. Three coupé versions with Touring coachwork took the first three places in a 24-hour race at Pescara, and another came fourth in the 1937 Mille Miglia. Alfa successes in this fabulous event were legion. With the exception of 1931, one type of Alfa or another won it outright from 1928 to 1938. Most notable were the 8C 2900As, which took the first three places in 1936.

By the end of the decade, though, production vehicles were leaving the factory in a trickle. Under Mussolini's direction, the main thrust was to match the German cars on the Grand Prix circuit. Jano's last creation in 1937, before leaving under a cloud for Lancia and later Ferrari, was the twin-supercharged V12. It produced some 430bhp, which was more than the rear axle could cope with, and in the wake of this failure, Alfa lost the services of the man who had been perhaps their greatest asset. Other engine designs were pressed into service, including the 3-litre V16 which managed second place in the 1938 Italian Grand Prix. The wide-angle V12 engine was tried in the Tipo 162, and the mid-mounted flat-twelve in the all-independent suspension Tipo 512. Development of this highly promising car was stalled by the advent of the Second World War.

Before hostilities began, German victories on the Grand Prix circuit had become so predictable that the promoters introduced voiturette racing as an entertaining diversion. This class was quite hotly contested, but it proved to be the salvation of Alfa Romeo's morale, as they managed to find success with the Gioacchino Columbo-designed single-supercharger straight-eight Tipo 158 Alfettas.

As well as the four-door touring car bodies, Alfa Romeo's welded sheet steel box-section chassis could be specified with coach-built bodies bordering on the outrageous. This fabulous 8C 2900B coupé of 1937 was built by Touring Superleggera of Milan. Just ten were made.

Nicola Romeo died on 15 August 1938, and therefore missed the devastation of the Portello factory. Its 8,500-strong workforce had contributed to the Italian war effort, and as a consequence it was a natural target for Allied bombing raids. The factory was hit twice in 1943 and again in 1944. Despite this apparently major setback, production of aero and marine engines was maintained, and the racing programme resumed almost the moment hostilities had ceased, using pre-war 1.5-litre Tipo 158 Alfettas, uncovered from their hiding place. With their derivatives, the 159s, Alfa Romeo achieved complete supremacy in the post-war years, really up to 1951. They recorded twenty-five Grand Prix victories in the hands of Giuseppe Farina and Juan Manuel Fangio, who took the first two F1 World Championships in 1950 and 1951 respectively.

When the formula changed, and in the face of increasing success for Ferrari, Alfa Romeo withdrew from Formula One, not returning to that particular arena until 1970, when they provided V8 engines for the up-and-coming McLaren and March teams. A 3-litre flat-twelve engine was fitted to one of Graham Hill's Lolas in 1975, and was taken up by Brabham the following year. Brabham continued with Alfa flat-twelve and V12 engines for four seasons, until 1979, when Alfa Romeo themselves came out with the flat-twelve powered Tipo 177. For 1980 hopes rested on the V12 Tipo 179 and 180, raced by Bruno Giacomelli, Vittorio Brambilla and Andrea de Cesaris; the unfortunate Patrick Depailler lost his life in a test session at Hockenheim. By the year's end Giacomelli was able to qualify his car on pole position, indicating its clear potential. However, in the five subsequent years Alfa Romeo's Formula One effort enjoyed little success. Even in the hands of Mario Andretti there was no joy in 1981, largely because the team was unprepared at the beginning of the season for the ban on sliding skirts as aerodynamic aids, and experiments with suspension systems to make up the deficit proved unsuccessful. Highlights in later years included the reappearance of the Tipo 179 in 1982 with a 1.5-litre turbocharged V8, and de Cesaris leading the Belgian Grand Prix at Spa for eighteen laps in 1983. Ricardo Patrese and Eddie Cheever drove Tipo 184Ts in 1984 with the cars resplendent in the bright green livery of Benetton fashions, but they were seldom on the pace, and even managed to collide on the first lap at Kyalami in 1985. Under something of a cloud, Alfa Romeo withdrew from Formula One at the end of the year, although their V8 engines powered the private Osella team from 1986 to 1988.

Any motor manufacturer who participates in motor sport, whether as a publicity-seeking exercise or to 'improve the breed', has its reputation bolstered to a great extent, provided there are at least some successes. If the 1980s represented something of a nadir in Alfa Romeo's sporting heritage, its credibility as a maker of production-line models was tarnished at the same time by the rust crisis. This affected not only Alfa Romeo and its rivals at Lancia and Fiat, but also much of the world's automobile industry. It is alleged to have begun during the 1970s, a legacy of importing and using cheap but contaminated steel.

Perhaps what carried the company through these troubled times was the enduring support of a core of enthusiasts who continued to buy and enjoy Alfa products. These people remembered Alfa Romeo's red-blooded days of the late 1920s and early '30s, and the post-war Grand Prix domination. Younger people recalled the heady days of the 1960s, when the little GT-Z coupés and later the Giulia Sprint GT and its GTA derivatives were invincible in their classes in sports and touring car races.

Alfa Romeo's participation in motor sport has been somewhat episodic, seeming to ebb and flow with the waves of corporate enthusiasm. In sports car racing, things never seemed quite so gloomy as during the 1979–85 Formula One débâcle, although perhaps Alfa Romeo lacked the consistency, commitment and even the administrative aptitude of Ferrari. Back in 1952 Alfa Romeo spent two seasons campaigning the purposeful Colli-bodied 6C 3000CM Disco Volante sports cars, which Fangio drove to second place in the 1953 Mille Miglia and Le Mans events. The graduation to mass production in the early 1950s allowed many more owners to go racing in events like the Mille Miglia, the Giro d'Italia and the Carrera Messicana. They campaigned with their 1900 saloons, Bertone-styled Giulietta Sprints and SSs, and Zagato-bodied SZ coupés, and there were very many successes at this level.

The Alfa management grew serious again in 1964 and bought Dr Carlo Chiti's Autodelta Racing Team; this formed the basis for virtually all works competition activity until 1985. Most of Autodelta's victories were achieved in the mid-to-late '60s with the Giulia GT-Z coupés in grand touring car racing, and GTA coupés in production car events. The car that contested the sports-prototype category was the fragile Tipo 33, first raced in 1967, with almost no success whatsoever. It seemed a pity that Autodelta had abandoned the TZ2 and GTA projects to concentrate on the sports-prototype category. The mid-engined 2-litre V8 held considerable potential, however, and Tipo 33/2s finished well up in events like the BOAC 500km at Brands Hatch, the Targa Florio and Le Mans in 1968. Not until 1971 did reliability and performance improve sufficiently for the 3-litre V8-engined 33/3 to notch up some meaningful successes, including victory in the BOAC race and the Targa Florio. With sports car racing dominated by Ferrari and then Matra over the following two seasons, Autodelta had to be content with just a handful of placings. With very little works opposition, the flat-twelve engined 33/TT 12 had more or less everything its own way in 1975 and again in 1977, but without top class competition the World Championship victories were rather hollow.

Meanwhile, production carried on apace with the Giulia 105-series saloons replacing the 750- and 101-series Giuliettas as bread-and-butter models in 1962; launched in 1966, the 105-series Pininfarina Spider epitomised wind-in-the-hair freedom for a newly wealthy generation. By this time production facilities had moved from Portello, restricted as it was from further development by a housing estate, to a new factory complex at Arese on the outskirts of Milan. The company's Balocco test track near Turin was opened the following year in 1964. By 1969 annual production had reached

The Alfasud was in production from 1971 to 1988.

100,000. At this point the more sedate 1750 and 2000cc Berlinas superseded the fluted and scalloped Giulia saloons in 1967, only to be replaced themselves by the Alfetta range in 1972. There has always been a period of up to two years' overlap on the production lines, where the incoming and outgoing models are produced side by side. Attracted by government incentives and the prospect of cheap labour, Alfa Romeo commenced production of the Alfasud at its Pomigliano d'Arco commercial vehicle plant near Naples. Despite appalling labour relations problems at Pomigliano, the Sprint derivative of this excellent little car was still being built in 1988. The Alfasud's successor, the 33, was also made there, although Pininfarina built the Sportwagon estate version.

However, despite its undoubted qualities, the Alfasud had cost Alfa Romeo dearly, and an attempt to head off financial disaster was made in 1980 by embarking on a joint venture with Nissan to produce an Alfasud-powered Nissan Cherry, marketed as the Arna. It was not a success in sales terms, and the venture ended. At Arese the Alfetta range was augmented in 1978 by the new Giulietta saloon, and by the Giugiaro-styled GTV6 coupé in 1980, which used the six-cylinder engine from the large Alfa 6 saloon, and enjoyed consistent success in European touring car racing in the early 1980s.

In 1987 the company was acquired by Fiat, after a take-over battle with Ford, from the state-owned Finmeccanica group, joining Lancia under the protective Fiat umbrella. That was possibly the most momentous occasion in Alfa Romeo history. With its future now relatively secure, Alfa began to rise with confidence from its lowly position in the automotive marketplace of the 1980s. It took a decade to achieve, but in 1997 its fabulous new 156 had scooped the coveted Car of the Year trophy.

Chapter Two
Mass Production

It's no exaggeration to say that Alfa Romeo virtually invented the sporting saloon in 1950 with the 1900, its first foray into mass production. Comparatively lightweight yet able to seat five people, the 1900 was the precursor of the even more popular Giulietta range that brought sports car motoring to the general public.

Alfa's wealthier customers in the immediate post-war period could order refined coupés like this handsome Villa d'Este 6C 2500 with coachwork by Carrozzeria Touring of Milan.

Like the rest of Alfa's output after the Second World War, the Villa d'Este was assembled from components produced before the war. Its drivetrain came from the 6C 2500 line.

The coachwork on this 1947 6C 2500 cabriolet is typical of its day, and in this case it's by Carrozzeria Ghia. Like most luxury cars of the period, it was made in right-hand-drive, and in this case the bonnet has a central hinge arrangement.

The 1900 was Alfa Romeo's first commercial post-war design, and was introduced in 1950. Its four-cylinder engine continued the company's twin-cam tradition, but incorporated a one-piece cast-iron block and aluminium head, with output rated at between 90 and 115bhp according to state of tune. Between 1950 and 1953, 7,611 units were produced. The 1900 was advertised under the slogan 'The family car that wins races', based on successes in events as diverse as the Tour de France, Targa Florio, Stella Alpina, Coupé des Alpes and the Carrera Messicana.

This is a 1900 TI – which stood for Turismo Internazionale. The 1900's one-piece unit-construction or monocoque body was Alfa's first design to have no separate chassis, as well as being its first mass production vehicle. Company boss Dr Orazio Satta realised that a highly developed independent suspension system was no good if it wasn't fully maintained, so elected for a simpler well-located solid rear axle with a triangular link and radius arms. Double wishbones were used in front, and large-finned aluminium drum brakes retarded the 1900's progress.

The face that launched many thousands of Alfa Romeos: this is the standard 1900 Normale featuring the shield-shaped central air intake flanked by chromed side intakes. As you might expect, the 1900s are rather heavy and ponderous to drive by today's touring car standards, but the 1900 was nevertheless an excellent practical classic. A certain number of mechanical and hydraulic parts are compatible with the subsequent 750-, 101- and 102-series cars.

Alongside the Berlinas were a number of exclusive coachbuilt coupés like this elegant Touring Superleggera 1900 Super Sprint made between 1951 and 1953. While the 1900 saloons were made of steel, the much rarer coupés' coachwork was generally in aluminium. They were usually described as three-window or five-window models, depending on the number of side windows. As a general yardstick, the earlier coupés usually had smaller rear windows and side doors that overlapped the sill, which was a stylistic throwback to the 6C 2500. Later models often had larger rear windows and doors that ended at the sill panels.

The 1900 drivetrain also powered some of Alfa Romeo's sports cars of the period. This is a 1900 Sport Spider of 1954, of which only two roadsters and two coupés were built. The wraparound windscreen, the works' four-leaf clover – or quadrifoglio – competition decal, and twin exhaust pipes exiting the sill are prominent, so there's no mistaking the car's purpose. It is now on display at Alfa's museum at Arese.

When Alfa Romeo sought a successor to the Touring-styled Disco Volante sports car in 1954, they called on Nuccio Bertone. He came up with a series of astonishing coupés that were known as the BAT cars. Any connotations with Batman were purely coincidental, as this was an acronym of Berlina Aerodinamica Tecnica. Beneath the extravagant bodyshell of BAT 5 (the first one to be built) lies a regular 1900 chassis and mechanicals.

Examples of this 1900 Super Sport were made by Carrozzeria Boano between 1952 and 1955. Mario Boano's career as a coachbuilder began with Giovanni Farina – Pinin's brother – and for a time he taught at the Turin school of automobile bodybuilding. He broke away from the Farina brothers in 1934 to pursue a solo career, which brought him to Carrozzeria Ghia in 1943. Boano was responsible for the original renderings for such classics as the Lancia Aurelia B20 and the Giulietta Sprint, and with his son Paolo went on to set up the Fiat Styling Centre and design school in Turin, which would evolve into the current Centro Stile, whose creations flourish as the 1990s crop of Alfas and Fiats.

A sporting saloon it definitely is not. But the 1900 power plant also did duty in the Alfa Romeo AR51 Matta jeep, seen here pressed into service as a funicular railway engine. Presumably the priest is present to administer the last rites if the brakes should fail! Seriously, Matta jeeps actually participated in the 1952 Mille Miglia, winning the military vehicle class.

The next phase of Alfa's saloon car production was the Giulietta range, but it was preceded by the pretty Giulietta Sprint coupé, which was first announced in 1954 as the prize in an Italian government lottery. At this point, the Berlina was only at the prototype stage. The first series of Giulietta models – Berlina, coupé and Spider – are identified as 750-series, and modifications made in 1959 introduced the 101-series with certain cosmetic changes. When a 1600cc twin-cam replaced the 1300cc engine, the three ranges became known as Giulias. This is a 101-series Giulia Sprint.

This is a 101-series Giulietta Sprint stripped down for racing purposes, but the beauty of its Bertone styling is not compromised. These cars rapidly achieved an excellent reputation for speed and agile handling, and demand for them caused havoc at the small Bertone factory. The Giulietta did wonders for both Alfa Romeo and coachbuilder Bertone, consolidating their reputations as credible stylist and volume manufacturer. By the summer of 1955 a two-seater Giulietta Spider was launched, with styling by Pininfarina.

To provide greater performance and further improve on the Giulietta's competition successes, Veloce versions of both Sprint and Spider were introduced in 1956. Although the figures reveal only a small horsepower increase, the Veloces were much faster in practice. The improvements resulted from a higher compression ratio, higher lift cams, twin Weber carburettors instead of the single Solex, tubular exhaust manifolds, and a finned cast-aluminium sump containing a built-in oil cooler and surge baffles to aid temperature control.

The Giulietta Sprint coupé's front suspension used twin wishbones, coil springs and dampers, while at the rear were two radius arms and a triangulating link that held the solid rear axle well located, further supported by coil springs.

The Giulietta's main attractions in the 1950s were its relatively high performance (despite its 1300cc engine), excellent handling and ride allied to remarkably efficient drum brakes, which provide a delightfully intimate driving experience not so far removed from more recent models.

The four-door Giulietta Berlina – 750C – was introduced in 1955, with styling for its steel body done in-house. It was produced at the Portello factory, and bodyshells can be seen going down the line mounted on heavy duty trolleys, while the operators grind off excess metal from the roof seams. The panels which combined to form the Giulietta's body-chassis unit were welded together in a series of jigs mounted on wheeled dais that travelled along a track, which was itself elevated on a raised wooden platform.

Suspended from an aerial gantry, Giulietta Berlina bodyshells are hosed down by waterproofed operators after undulations in the metal have been filled in, prior to being painted.

The line of Giulietta Berlina shells nearing completion at the Portello factory, as chrome trim and lights are installed. The side-grille trim and side indicators on the front wing identify these cars as early 101-series TI models, dating from 1959.

With front wings and panels protected by sheets, the almost completed Giulietta Berlinas and Sprints on the overhead assembly lines have their engines and ancillaries installed. A row of engines and stack of propshafts await installation to the left of the picture, while a line of Spiders is finished off on the extreme left.

This is an early 101-series Giulietta TI, dating from 1959. The TI variant was introduced in 1957 and differed from the basic Berlina Normale in having a twin choke Solex carburettor in place of the single choke Solex, but didn't go as far as the Veloce models in increasing the performance significantly.

One of the chief criticisms levelled at the Giulietta Berlina was the lack of legroom in the rear cabin. However, it was quite adequate in the front, and the controls were pretty much state of the art. There was a column shift, which provided reasonably efficient gearchanges once the action was mastered, and an organ-type accelerator pedal. The clutch and brake were bottom-hinged, while the speedo was of the strip variety.

This is a 750-series Giulietta Berlina of about 1957, wearing Pirelli Cinturato tyres. The trim of the bumpers and side-grilles is much plainer than on later models, and there is a much smaller side-indicator on the front wing close to the A-pillar. It also has a piece of trim that serves as a bonnet handle, but this disappeared with the 101-series cars.

There were a number of variations on the Giulietta Berlina theme: 71 units of an estate car known as the Promiscua had their Giardinetta bodyshells built by the Colli brothers, who also made the fabulous 6C 3000CM racing coupés driven by Fangio among others. Then there was a long-wheelbase limousine also made by Colli. But although the Berlina was the bread-and-butter model, it was not made in right-hand-drive – like this car – until 1961, when the Normale and TI saloons were uprated along with the rest of the Giulietta range.

The revised Giulietta TI made its debut at the Frankfurt Show in 1959. It featured recessed headlights and rubber inserts in the bumper over-riders. This beautiful blue example belonging to John Brittan was first registered in 1963 and shows the redesigned shield and far broader side-grilles, intended to provide better cooling for the new 1600cc engine.

The badging on the bootlid of this Giulietta identifies it as a TI model, and it displays the larger rear light clusters of the facelifted cars. While the 101-series Spider and Sprint became Giulia models from 1962, the saloons never relinquished their identity as Giuliettas. This makes life easier, as they can be distinguished from their successors more simply than their sporting siblings.

One important benefit that the post-1959 cars had was a floor shift, although there were still only four forward gears. This right-hand-drive Giulietta TI has a Bakelite steering wheel with chrome horn ring, strip speedometer and clear dials, and easily accessible switchgear. The manual choke levers are visible below the ignition key fob.

The Giulietta TI's four cylinder, Solex twin choke carb, twin-cam engine developed 74bhp at 6200rpm, while the lower compression Normale produced 62bhp at 6000rpm. In 1961 the engine was given a six-bladed white plastic cooling fan to replace the four-bladed aluminium one.

Here's a variation on the coupé theme. With its rotund, lightweight bodyshell designed and built by Zagato between 1959 and 1962, the Giulietta SZ was a very successful competition car in Italy and continental Europe. It was powered by the 1300 Veloce engine, and its cabin was quite austerely furnished.

This little cutie was a front-wheel-drive prototype built in 1960. Designated the Tipo 103, it was powered by the smallest configuration of the twin-cam engine at just 896cc. More radically, the four-speed gearbox was constructed as part of the engine, and the whole assembly was transverse-mounted. Unusually for 1960, the camshafts were driven by synthetic rubber belts, while the generator used twisted and intermediate pulleys. The cross-flow engine delivered 52bhp at 5500rpm. The spare wheel was housed under the rear seat, which may have posed some practical difficulties. There were some stylistic similarities with the much bigger 102-series 2000 saloon and the forthcoming Giulia TI. But the project was stymied not only because of the costs of erecting another production line at the new Arese plant, but also because of Alfa Romeo's pact with Renault to avoid direct competition with cars like the R8. The single 103-series prototype was built, along with three engines. It was re-examined when the Alfasud project was first mooted, but by the late 1960s much of its design was obsolete. The front-drive concept would emerge again in 1972, but in a very different way.

The 102-series 2000 models were introduced as replacements for the 1900. As with the Giulietta, there were three models: Berlina, Spider and Sprint. The saloon and sports car appeared first in 1958 and the coupé followed in 1960. The Berlina was an in-house design, the Spider was styled by Touring-of-Milan, and the Sprint was by Bertone. The 102-series 2000 engine had a certain amount in common with the Giulietta in the valve department, but like the 1900 it had an iron block, individual cam covers and a cam-driven distributor. It shared its five-speed transmission with the much more exotic 101-series Giulietta SS and SZ coupés.

These cars were significantly larger than the Giulietta range, and the Berlina's cavernous boot in particular provided generous luggage space. They remained in production until 1962, although only 2,947 examples were built. The Spider was the most popular with 3,443 units made; this is unusual because normally the saloons were the most numerous and were the basis for most Alfa Romeo production cars. Although it was probably the company's least successful range, the 2000's strength lay in its smoothness and relaxed high-speed cruising, and both the engine and chassis developed reputations for durability. However, it was underpowered, and inevitably failed to match the Giulietta's capacity to entertain.

The main styling cues of the 2000 Berlina were carried through to its successor, the 2600 – or 106-series – model. It differed in appearance from the 2000 in having a second set of driving lamps mounted in the plainer radiator grille, and the Farina-esque tail fins were pruned back to produce a flatter Kamm-tail rear end. They were powered by a new engine, a high-revving straight-six twin-cam unit that enabled the 2600 Berlina to reach 108.3mph and make 0–60mph in 13 seconds. While its sister cars, the Sprint and Spider, used three twin choke Solex carbs, the Berlina made do with a pair of twin choke Solex 32PATA carbs. The saloon also had a slightly lower compression ratio of 8.5:1.

The 106-series can be seen as a deliberate return to the more refined sports-touring market that Alfa Romeo had largely abandoned when it struck out with the smaller Giulietta range of four-cylinder models. Here, a 2600 Berlina and a Touring-bodied 2600 Spider are being put through their paces during a club meeting at Kyalami circuit in South Africa in 1974.

Although the 2600 Berlina was technically interesting and aesthetically pleasing in an early '60s context, it failed to win many customers in the showroom. As tastes changed in favour of a plainer slab-sided look, the chrome strips that had adorned the flanks of the earlier 2600s were deleted by 1965. Just 2,092 Berlinas, 6,999 Sprints and 2,255 Spiders were built, and production ceased in 1968.

A small number of 2600s were constructed by specialist coachbuilders, and this is one of fifty Berlinas made by Osi in 1965. Its rather flatter lines tend to elongate the appearance, and it has a more contemporary look than its factory-produced sibling.

The cabin of the 2600 Berlina was an appropriately opulent environment, and it was naturally far more spacious, comfortable and tranquil than those of either the coupé or Spider, and there was substantial accommodation in the rear as well. Interior layout progressed from a full bench seat to split bench to separate seats with cloth inserts by 1965. It was normal for left-hand-drive cars to have a column gear shift, and right-hand-drive-cars to have a floor shift, but paradoxically this left-hooker has a floor shift. The driver's seat has comprehensive adjustment, while the brake and clutch pedal are floor-hinged.

The 2600 Sprint came out in 1962 and it differed from the old 2000 Sprint only in minor differences of trim and badging. Its 145bhp six-cylinder motor fitted remarkably well into an engine bay originally intended for the smaller four-cylinder unit, and in fact its Giugiaro-styled proportions seem to cry out for the bigger power unit. Mechanically, the only other revision was the adoption of disc brakes all round. The cabin interior featured electric windows, although they proved to be extremely slow in operation. Just under seven thousand 2600 Sprints were built before production ended in 1966.

Like the coachbuilt sports-touring machines on which Alfa Romeo's pre- and immediate post-war reputation had been built, the 2600 provided an excellent basis for the specialist coachbuilders to display their skills. This is Zagato's offering on the Sprint theme, and it is typically extravagant in some of its detailing, such as the treatment of the radiator grille. The SZ was first seen on the Zagato stand at the Turin Show in 1963, and featured on the Alfa stand the following two years. Zagato produced a mere 105 2600 SZs in total.

This Touring-bodied 2000 Sprint coupé of 1960 is altogether more restrained, and its two-plus-two proportions are perhaps compromised by its relatively short wheelbase and lengthy rear overhang. There is also a curious styling quirk in the concave rear window. Nevertheless, it gives the impression of an elegant and expensive vehicle, which was always Touring's forte.

This elegant beauty is a Giulietta Sprint Speciale, introduced in 1959 and based on 101-series mechanicals and floorpan with steel bodywork (and aluminium bootlid) by Bertone. Some 1,350 units were made up to 1962, succeeded by a further 1,400 units that were identical externally but ran with Veloce-spec 1600cc engines and were known as the Giulia SS. They can be regarded as the productionised version of Bertone's earlier BAT cars, and were virtually hand-built with well-finished, albeit cramped cockpits. Note the perspex fly-screen mounted ahead of the windscreen.

Chapter Three
From Giulietta to Giulia

In 1961 the 100,000th Giulietta rolled off the Portello assembly line. But waiting in the wings was the new Giulia series which would be produced in a new factory in the northern suburbs of Milan. The new range included competition cars like the Giulia TI/Super and GTA which would bring the company numerous racing successes.

Alfa Romeo production shifted from Portello to a new plant at Arese to the north-west of Milan, and in 1962 the new 105-series cars started life there. This photograph, taken a decade later, shows the administration and museum block in the right foreground.

The 105-series range consisted of the saloon, coupé and sports car variants, and began with the Giulia TI saloon, pictured here, launched in 1962. The generic Berlina name-tag was shelved for the time being. The TI's floorpan and running gear formed the basis for the subsequent Sprint GT and Duetto Spider models, and in the sense that the Spider was not replaced until 1993, its 105-series componentry endured for a very long time by motor industry standards.

The Giulia TI 1600 was powered by the 1570cc Alfa Romeo twin-cam engine. Almost immediately, a competition derivative appeared known as the TI/Super, which could be specified with perspex rear windows, twin Weber carburettors and alloy wheels. During the next few years it saw action on circuits the world over, usually locked in combat with the ubiquitous Lotus Cortinas.

This pristine Giulia TI/Super was discovered by its current owner Richard Everton in daily use as a shopping vehicle in Provence, and was imported to the UK where it was repainted by Alfa restoration specialist Mike Spenceley. The inner pair of headlights, normally the main beam ones, were frequently removed to aid engine cooling, and the model's unique badging and green cloverleaf sticker are prominent.

One of the foremost exponents of Alfa Romeo racing in the 1960s was Andrea De Adamich, and his career encompassed saloons, touring cars, sports prototypes but ended in Formula One when he suffered a broken ankle in the notorious start line pile-up at Silverstone at the 1973 British Grand Prix. Driven by De Adamich and Arcioni, and entered by Scuderia Jolly Club, this TI/Super was the outright winner of the annual Monza Four Hours race, held on 19 March 1965.

The interior of the TI/Super features leather-trimmed bucket seats and a simplified dashboard. Window winders in the rear doors show that this car has glass panes.

The engine bay of the Giulia TI/Super is a relatively uncomplicated space, providing easy access to the 1570cc twin-cam engine. The twin sidedraught Weber carbs breathe through characteristically elephantine trunking and air filter. The single brake servo is to the rear.

National owners' clubs have at least one major annual get-together, and this is the 1998 Dutch meeting at the Assen race circuit in northern Holland. The Dutch have a particularly soft spot for the Giulia saloon, and hold a race series specially for competitors in these cars, known as the Squadra Blanca. I even discovered a restorer in north Amsterdam who specialised in restoring Giulia Supers he'd brought in from Italy.

Giulia TI/Supers – or regular Giulia saloons got up to look like them – compete regularly in classic events that replicate the European Touring Car series of the 1960s and '70s, and bring together a host of racers from those halcyon days. Here, a Giulia Super mixes it with one of its traditional adversaries, the BMW 1800, together with a Giulia Sprint GT, a Mk II Jaguar, a Mercedes Benz and sundry other Alfas and Mini Coopers at the Coys International meeting at Silverstone in 1993.

A pair of Giulia TI/Supers high on the banking at Monza during the Coppa Inter Europa held on 25 October 1964. No. 65 Riccardo di Bono leads team-mate and former works Ferrari and Maserati driver Gino Munaron. The finishing order was reversed as di Bono took third place and Munaron came second. Munaron also shared a TI/Super with de Adamich to finish fifth in the Spa 24 Hours in 1964.

Another works Giulia TI/Super on the pits straight at Monza during the Coppa Inter Europa in October 1964. The cars are equipped with fly screens intended to collect oil, bugs and debris that would otherwise obscure the windscreen. Bearing in mind the cracks in the concrete surface, the suspension must have taken quite a pounding.

So this is how they did things in the swinging Sixties! A quick romp in the meadow and then off for a burn-up in the Alfa. Clearly the Giulia was some car if it could inspire such abandoned behaviour. More to the point, this is a Giulia Super, introduced in 1965 and the model that became the mainstay of Alfa Romeo production for the next decade.

Fans of the cult movie *The Italian Job* will recall Michael Caine's trio of Mini Coopers running rings around the inept Milanese police in their 1600cc Giulia Supers. I may be biased, but in reality I bet the Alfas would have given the Minis a better run for their money than they did in the film! Pictured is an authentic Milanese police car of 1969.

The driving position of the Giulia Super fitted into the category often described as 'Italianate', meaning that tall people with long legs had to adopt a knees-splayed posture because of the relationship of the steering wheel to the pedals and the seat. On the other hand, the gear lever was well placed for those marvellous 'knife-through-butter' gear shifts that were never really bettered. This car belongs to Richard Everton and has a wood-rim wheel.

Another area where the saloons naturally score over their sporting siblings is the luggage boot, which is invariably as capacious as that of any of the competition. The Giulia Super's jack and modest toolkit is stowed against the rear bulkhead with a butterfly nut, and the spare wheel lies in a well in the boot floor.

Door furniture was pretty basic in the Giulia 1300 TI saloons, although the winders and door handles worked well enough. The quarter-light window was an efficient way of ducting fresh air into the cabin. On the 1600 Super, this window was opened with a separate winder rather than the simple catch seen here.

The Giulia 1300 TI had a bakelite steering wheel and a less ostentatious dashboard and instrumentation than the 1600 Super, but the vinyl-covered seats and interior trim were of similar quality.

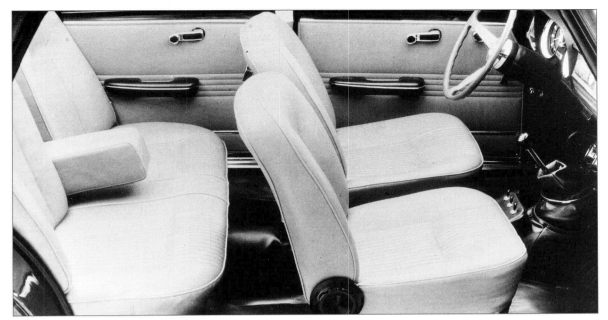

Interior of the Giulia Super in 1972, showing how the rear seats were nicely sculpted, with arm rest down. The front seats could be reclined fully by the twist knob. Leg room was adequate in the rear, so long as the front seats were well forward.

The Giulia 1600 TI was joined by the 1300 TI in 1964, posed here overlooking the beautiful Lake Orta. Apart from variations in the interior trim, the most prominent difference between the 1300 model and its bigger capacity sister was the single headlight façade.

Surprisingly, the boxy shape of the Giulia saloon was remarkably effective aerodynamically. It was the product of wind-tunnel testing, and the steeply raked windscreen, cut-off Kamm tail and attractive flutings along the car's flanks and roofline produced a drag coefficient figure of cd 0.33. That's far superior to that of a host of more modern and supposedly aerodynamic coupés, including its own Bertone-designed siblings. This is a 1969 1300TI, still with single headlights, but a revised grille treatment.

One of the attractions of these classic Alfa Romeos is that the engines are readily interchangeable because the mounting points are compatible throughout the range. I ran this 1971 Giulia 1300 TI during the latter half of the 1980s when working on *Restoring Classic Cars* magazine, and switched to progressively larger capacity power units in a quest for greater performance without courting the questionable reliability of performance tuning. By degrees it moved up from 1600 to 1750 and finally ended up with a 2-litre engine. The final drive was also swapped for a limited slip differential from a later 2000 Berlina.

Giulia saloons have been known to take outright honours at National Alfa Day concours events. This is the Giulia Super in its final form in 1976 when it was known as the Nuova: the flutings in the bonnet and boot have been ironed out, and the grille arrangement of the 2000 Berlina adopted. Demonstrating a degree of corporate versatility, this particular car is powered by a 1760cc Perkins diesel engine, installed in 1976 as an economy measure. It produced just 55bhp, and some 6,500 units were made with this engine.

Between 1963 and 1967 Carrozzeria Zagato built 112 of these purposeful little coupés, known as the GTZ, or Giulia Tubolare Zagato. The bodies were constructed in aluminium panels around a rigid tubular steel frame. Although they were intended for competition use, a number of GTZs were used as road cars, and they ran with the 1600 TI/Super engine and 45DCOE Weber carbs. The engine had to be canted over at such an angle that several components had to be specially made, such as the sump, gearbox bell-housing and exhaust manifolds.

There's no better example of a Kamm tail than the rear end of a GTZ. Snapped at the AROC annual get-together in 1992, this pretty Giulia GTZ demonstrates the severely truncated tail that resulted from aerodynamic experiments carried out by Professor Wunibald Kamm. He discovered that such a configuration reduced drag and improved airflow, and the lessons were applied with great success to the Giulia saloon. Just beyond the GTZ in the picture is an 8C 2300 Tourer.

The final twelve cars in the GTZ production run were substantially more aggressive in appearance, and were clad in fibreglass body shells. Known as TZ2s, they were equipped with the GTA's twin-plug cylinder head and wider 13-inch wheels. This is the TZ2 of Teodoro Zeccoli and 'Geki' Russo on their way to third place in the 1600cc class on the 1966 Targa Florio, the kind of endurance race that the works-backed Autodelta team specialised in.

The coupé model in the 105-series range was the Giulia Sprint GT, introduced in 1963. It was joined by the externally identical Veloce model shown here in 1965, which despite its twin Weber carbs and tapered inlet porting managed only a 3bhp gain over the standard Sprint GT.

The sporting derivative of the Giulia Sprint GT was the GTA – the 'A' stood for alleggerita, or lightened – and all the exterior panels were in lightweight aluminium. Only the floorpan, inner shell and sills were in steel. The minimal interior trim and upholstery were also lightened in a bid to improve the power-to-weight ratio, all contriving to reduce the weight of the standard steel-bodied car by 500lb. The GTA also parted company with regular Giulia Sprint GTs in having a twin-plug cylinder head, 10.5:1 high compression head and twin 45mm DCOE Weber carbs. The factory produced 500 GTAs between 1965 and 1967, and this is one of fifty units made in right-hand-drive.

In the hands of drivers like Andrea de Adamich (pictured here on a T33 sports prototype), Ignazio Giunti, Nanni Galli and Spartaco Dini, the Autodelta GTAs were spectacular to watch and extremely successful in touring car races all around the world as well as in Europe. The legendary Jochen Rindt was drafted into the squad to win the 1966 Sebring Four-Hour race for Autodelta, and my enthusiasm for these cars was kindled at Snetterton the same year, watching Rindt and de Adamich's thrilling battles with the Lotus Cortinas of Jackie Stewart and Sir John Whitmore.

Another derivative of the Giulia Sprint GT was the 1300 GTA Junior, which used the 1290cc short stroke twin-cam engine with twin-plug head and close ratio gearbox. It was first seen in 1968, and although records state that 447 were built, the true figure may well be closer to a thousand. The interior was less austere than in the 1600 GTA, with more attention to soundproofing. This one carries Alfa's serpent and cloverleaf decals.

A 105-series coupé bodyshell is hoisted aloft by an operator on the Arese production line. The twin headlight shell, plus the Berlina bodies in the background, indicates that these are destined to be 1750 models, which first appeared in 1967 alongside the Sprint GT and Giulia Super.

Giulia saloon and coupé bodyshells are elevated on a maze of overhead gantries at the Arese plant prior to having their drivetrains and suspension fitted.

The distinctive flutes along the roofline, the flanks, bonnet and boot of the Giulia Super were erased on the 1750 Berlina of 1967 that was intended to replace it. Bertone was engaged to provide both the front and rear end with a more sober treatment to create a more up-market image, although the Berlina used many of the same pressings as the Super. The twin Carello halogen headlights provided exceptionally good illumination, and could be adjusted if the car was heavily laden. As on all 105-series saloons and coupés, the windscreen wipers perform an idiosyncratic cross-arms routine when in operation.

Boot capacity was slightly better in the 1750 Berlina than in the Super. My formative years as an Alfa Romeo enthusiast were spent in one of these Berlinas, and it was an extremely practical car. I once moved the entire contents of my flat in it, and spent long hours cruising the motorways and A-roads between London and Scotland, where it loped along effortlessly at a steady 100mph.

The interior of the 1750 Berlina's cabin is not so different from that of the Giulia Super. But everything is that bit more luxurious, with carpet instead of rubber matting, more integrated armrests and door-pulls, new locks, more comprehensive instrumentation and a characteristic binnacle housing the speedo and rev-counter. The gearstick still projected back at the driver at a 45-degree angle. The 1750's successor, the 2000 Berlina, had seats upholstered in cloth, and was equipped with headrests back and front.

The British Alfa Romeo Owners Club meets annually at Stanford Hall in Northamptonshire, and one of the side attractions is the gymkhana. This is Chris Taylor's 1750 Berlina in the midst of the slalom course in 1990. This car has a roll-cage installed and lowered suspension for competing in the Alfa Romeo club racing series. The set-up consisted of coil springs and dampers, wishbones and anti-roll bar at the front, with a live axle, coil springs and dampers, trailing arms and an anti-roll bar at the rear.

In 1972 the 1750 Berlina and its coupé brethren were phased out and replaced by the 2000 range. The elegant model is present to emphasise the opulent trend of Alfa's saloon range. The 2000 Berlina provided similar dynamics to its predecessor's in terms of handling, but instead of the free-revving 1750 engine, the 2-litre lump was a more deep-throated and torquier proposition. External changes were limited to badging, the larger Alfa grille, rear light clusters, side lights and side indicators. Hubcaps that covered the wheel nuts were replaced by central dust covers.

Rear three-quarter view of the 2000 Berlina showing off the flattened bootlid and tail panel. The wraparound rear screen has also gone, and the appearance is considerably less charming than that of the Giulia Super, which was still being produced alongside it, although there was a substantial gain in performance terms from the bigger engine. The five-speed gearbox now drove through a limited slip differential. While in European markets the 2000 range continued to be 105-series cars, the chassis prefix on US cars was changed and they became 115-series models accordingly.

The dashboard of the 2000 Berlina was perhaps less exuberant than the 1750's, with the revised speedo and rev-counter flanking a column of warning lights housed in a more straightforward panel ahead of the driver. The central console from which the gear lever protrudes is now clad in fake wood veneer, as are the auxiliary dials in the dash. The rubber button prominent in the footwell to the left is for the windscreen washer. The two-speed wiper switch was located, oddly, on the centre console behind the gear lever.

The 2000 Berlina was also available with German ZF three-speed automatic transmission, and while the main controls are identical to those of the manual gearbox model, apart from the shift and pedals, there is less evidence of the fake woodwork on the transmission tunnel. All Alfa saloons score well in having a footrest to the left of the clutch.

Yes, it's Stirling Moss, and like one or two other racing drivers of the early 1970s – including Peter Revson, Mike Hailwood, and Andrea de Adamich – he drove a Montreal. Originally designed by Bertone as a two-seater prototype based on the Giulia 105-series platform and running gear for the 1967 Montreal World Trade Fair, the eponymous coupé graduated from 1600cc to a full-house 2.6-litre V8 when it was productionised in 1970. Moss was quoted in the Italian press as being very impressed by the car's amazing performance and surprised that it was powered by only a 2.8-litre engine. It may have been of relatively modest capacity, but it was derived from the T33 sports prototype.

The 1600 Junior Z was another creation by Zagato, based on the Sprint GT floorpan and running gear. It was built in 1300cc form between 1970 and 1972, with 1,108 units made, and as a 1600 between 1972 and 1975, with 402 cars built. The later cars were slightly longer, with different rear lights, and the fuel filler swapped sides. Contrary to Zagato's normal practice of using aluminium, the Junior Z was made in steel. Perhaps its most distinctive feature is the full-width plexiglass front panel, which has the Alfa shield shape cut out of it.

Racing improves the breed, no question of that, and Alfa's 105-series cars were a case in point. The Giulia GTA coupé metamorphosed into the GTA SA – the SA stood for sovralimentazione, or supercharged – and they showed considerable potential during 1967 and 1968. This is one of two Autodelta cars that I snapped in the Snetterton paddock during the annual European 500km Touring Car thrash on Good Friday 1967, where both drivers, Galli and Roberto Bussinello, retired, one shunted off the start-line and the other punted off on lap two. The GTA SA was abandoned when regulations for the European series changed from Group 5 to Group 2, which would have meant Autodelta building a thousand supercharged cars.

In 1970 the fully developed GTA appeared, known as the GTAm. According to the official homologation papers, 'Am' is an abbreviation for American because the car used the Spica mechanical fuel injection-system fitted to North American spec production cars, which implied the 1750 range. However, the Autodelta GTAs had been running fuel injection since 1968. The 1750 GTAm was listed as a production car for 1970, so building a thousand examples presented no problem. Autodelta simply took forty of these and turned them into racing cars. The engines were bored out to 1985cc and there were four valves per cylinder, but only a single spark plug for each. The racers differed visibly from the road version in the enormous flared wheel arches that accommodated the Tipo 33 magnesium alloy wheels and wide slick tyres. The steel wheel arches were simply cut away and bulbous fibreglass ones bonded in place.

The GTAm scored sufficient victories in the 2-litre class to scoop the title outright in the 1970 European Touring Car Series which demonstrates the pace and reliability these Alfas were capable of. Serving to reinforce the point were Nanni Galli, who finished the 1970 Tourist Trophy in fourth place, and Gian-Luigi Picchi, sixth, in two more GTAms. Picchi's car, pictured here, is cocking a front wheel in characteristic GTA fashion, the legacy of stiff front end and softer rear set-up. These were still the days when the stars of F1 and international rallying turned their hands to anything, and the entry list for this race included Jackie Stewart, Rolf Stommelen, Helmut Marko and Rauno Aaltonen, as well as the regular touring car stalwarts.

A pair of GTAms hunt down a BMW Alpina 3.0CSL in the sand dunes at Zandvoort in August 1970, where Gian-Luigi Picchi, left, took a class win. The following year Picchi tied for first place with Hezemans in the drivers' championship, having notched up a string of class wins in a 1300 GTA Junior. In 1972 the 1300 GTA Juniors swept the board – no fewer than nine of them raced in the TT at Silverstone that year.

While the 2000 Berlina provided enervating transport for the more mature motorist, its grand touring sibling, the 2000 GTV, was no less entertaining but was considerably less spacious. Those who wanted a quieter life could opt for either the 1600 or 1300 GT Junior, introduced in 1974 in the same shell as the 2000 GTV, but in single-headlight guise. The rakish lines were more or less unaltered from Giugiaro's original rendering for Bertone's Sprint GT, but the smaller capacity models lacked the 2000's more comprehensive rear light clusters which included reversing lights. As in the preceding model, the single reversing light was below the bumper in the middle of the rear valance. Badges on the C-pillar swapped the green cloverleaf emblem for Alfa's Visconti-derived serpentine trademark.

The engine bay of the 1600 GT Junior, showing the distinctive twin-cam cover on the 1570cc all-aluminium motor, the pair of brake servos and twin sidedraught 40DCOE Webers with cylindrical air filter. It delivered 125bhp at 6000rpm, and 116lb/ft of torque at 2800rpm. Top speed was 112mph at 5700rpm in top, and the 0–60mph time was 10.3 secs.

The interior of the 1600 GT Junior was not as palatial as the 2000 GTV, lacking the cloth seats, but it was fitted with the steeply dished wood-rim wheel and the interestingly laid out instrument binnacle. The wood-grain effect that afflicted the Berlina was also applied to the coupé's centre console.

CHAPTER FOUR
MASS APPEAL

If any Alfa saloon captured the imagination of the general public, it was the Alfasud. Its successor, the 33, was produced in the greatest numbers of any Alfa Romeo to date.

A pre-production Alfasud is given shakedown tests near a Moroccan oasis. Like most prototypes, the 'Sud is camouflaged by having extensions to front and rear ends. The wildlife was distinctly unimpressed.

This is the basic Alfasud as launched at the Turin Show in 1971. Its compact four-door body was styled by Giorgetto Giugiaro, and some would argue that the basic concept was never improved upon. It differed from all previous Alfas in that it was front-wheel-drive and powered by a small 1200cc four-cylinder flat-four engine. This configuration was chosen in order to keep the design as compact as possible. On its launch the 'Sud was considered radical, which it was for Alfa Romeo, but it certainly was a landmark car in terms of combining driving pleasure with modest size and practicality.

The new Alfasud was 14 inches shorter than the contemporary Giulia range and 450lb lighter than the Berlina. Its low centre of gravity and wide track ensured that its roadholding was in a different class – or perhaps that should be style – to its predecessors. In other words it was easy to drive quickly because its handling was neutral and it cornered as if on rails, and it gained the reputation overnight of being an excellent car to drive. This is an Alfasud L of 1974.

The Alfasud's flat-four engine was located far forward in the monocoque shell, and therefore, as in other front-wheel-drive cars like the BMC 1100 range, the space for occupants in the interior could be much greater. Seats were state-of-the-art for a mass-produced car, with headrests and reclining facility in the front.

The original dashboard layout and controls of the Alfasud were a model of simplicity, but in typical Italian style the driving position suited shorter drivers with longer arms. Alfa never seemed to learn the lesson about providing reach adjustment for the steering wheel. On right-hand-drive cars, the pedals were offset to the left to give clearance to the off-side front wheel arch. Apart from the choke and windscreen washer, all controls were operated by two stalks on the steering column: lights and indicators to the left and heater, wipers and horn to the right.

By 1975 Alfa was offering a practical estate-car version of the 'Sud, known as the Giardinetta. It was undeniably a useful vehicle, but while its underpinnings were unmistakably Alfasud, there were unfortunate stylistic connotations with the unloved Austin Allegro.

From the rear the Giardinetta looks like any other small estate, and when it was fully laden, much of the vivacity of the car from which it was derived would disappear. Mind you, at this point in its history the regular 'Sud's boot was hinged just below its rear window, and lacked a stay to prop it up so it had to rest on the rear window, which further impeded stowing the luggage, so there was an argument in favour of the Giardinetta in this respect.

If the original Alfasud was inadequate in any way, it was in its poor top-gear acceleration because of a lack of torque at low revs. This deficiency was redressed with the Alfasud ti that appeared in 1974, and was accomplished by raising the compression ratio slightly, from 8.8:1 to 9.0:1 and replacing the single-choke Solex of the standard car with a twin-choke Weber. Although the bhp figure was only lifted from 63 to 68, the improvement in performance was more pronounced than the figures suggest. The two-door 'Sud ti was equipped with a five-speed gearbox, and a rev-counter was provided in this model for the benefit of the press-on motorist – the 1186cc engine worked best when revved hard.

Alfa Romeo sought to reach as wide a market as possible for the Alfasud by introducing different versions, and this is the four-door 5M of 1976. It had the higher level of trim associated with its predecessor, the L model – as opposed to the base N version – but it was given the ti's five-speed gearbox. While the original Alfasud had a clear price advantage over anything offering comparable performance and practicality, such as the Citroën GS and VW Golf, this was gradually whittled away by successive price increases to the point where purchasing it was a decision for the heart rather than the head.

In 1977 the Alfasud ti (turismo internazionale) was given the 1300cc engine that powered its newly introduced sister model, the Sprint coupé. This longer-stroke engine, combined with the five-speed gearbox, made a big improvement in acceleration all the way through the rev range, as well as making for better economy. It was a more relaxed car to drive, as it was no longer absolutely necessary to row it along by the gear stick. The 1978 car pictured here has new-style alloy wheels and boot spoiler.

The 'Sud ti's dashboard was tidy, if somewhat basic, and although its seating and controls were originally considered acceptable and any deficiencies were overlooked because of its excellent handling and performance, by 1978 these aspects of its interior were beginning to be criticised as cheap and flimsy and its fittings as crude and poorly finished.

In 1980 Alfa Romeo was in the throes of applying serious facelifts in a bid to update all its model ranges, and the Alfasud was no exception. This is the back of a 1.3 ti, showing the matt-black wraparound plastic bumpers and revised tail-light clusters. The boot spoiler was complemented by an air-dam that formed the front valance.

Apart from the usual shopping bags and holdalls, the Alfasud range could also transport skiing equipment, thanks to the adaptation of the rear seat so that longer items could be stowed lengthways within the car. The bootlid was still hinged below rear window level though, and in this respect it lagged behind all its competitors.

In 1981 the issue of stowing and carrying luggage in the Alfasud was addressed by the introduction of a proper hatchback, demonstrated here in a 'Sud ti. Whether or not you'd want to travel far with a weighty outboard and sundry other boating apparatus is another matter, but with a fold-down rear seat at least that was now possible. And the rear window could now be kept clean with the new wash-wipe system.

The external facelift also carried with it an updated Alfasud dashboard. All the controls were fundamentally as before, but the overall look of the instrument panel brought things up to date. There's a gaping hole where you'd expect the hi-fi to be, but the switches above are for rear wash-wipe, rear fog lamp, hazard warning, and heated rear window, together with a choke warning light and digital clock.

Following the example of the Giulia coupé range, which contained the 1600 and 1300 GT Junior and 1300 GTA Junior during the early 1970s, the Alfasud range also included a Junior model, identified by additional stripes and badges that were intended to impress the younger buyer. The plastic front air dam and sill trim that came with the 1980 facelift can be seen on this 1982 Junior.

Sporting yet another variation on the alloy wheel theme, sometimes referred to as 'telephone dials', this Alfasud Ti Green Cloverleaf of 1982 has the three-door hatchback body and is powered by a twin-carb version of the 1500cc engine that was first used in the Sprint coupé in 1979.

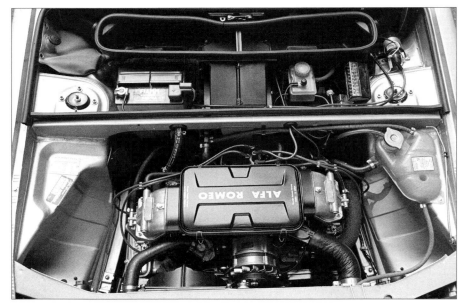

The engine bay of the Alfasud Ti Green Cloverleaf, showing the typical 'Sud bulkhead separating the motor from the battery, heater, fuse box and fluid reservoir. The engine itself was always mounted in this position, ahead of the suspension turrets, with the radiator's reservoir mounted above the left-hand inner wheel arch. The 'Sud's cast-iron flat-four, with its characteristic boxer rasp, was lumpy at low revs in this 1490cc twin-Weber carb format, but power delivery was smooth and refined at higher revs.

Cutaway illustration by Bruno Betti of the anatomy of the 'Sud's flat-four engine, showing the twin downdraught carbs, concave piston tops and single row of valves in the 'roof' of the combustion chamber, and single camshafts at either side of the unit. Having the weight distribution low down in the car made for a lower centre of gravity than could be obtained with a regular in-line engine, although its forward location ensured that the car had a frontal weight bias. It was none the less a compact and efficient design.

This twin headlight Alfasud Ti 1.5 of 1979 was the concours winner of the UK Alfa Romeo Owners Club's Surrey section when it belonged to Ian Brookfield in 1988. The car is equipped with racing seats and harnesses, a full roll cage and lowered suspension with Revolution alloy wheels.

Of all the corners on British race circuits, Snetterton's Russell bend is one of the most demanding. You approach the complex from the high-speed right-hander that is Coram and stand on the brakes while fiercely down-shifting to second gear, and simultaneously wrestling with the steering to pitch the car into the swift right-left, meanwhile ensuring that no one's out-braked you. Here Ian Brookfield's 'Sud succumbs to the gravel trap as he takes a short-cut across the apex during an AROC championship round.

It's that man Ian Brookfield again, and this time he's trying too hard on the exit of a bend at Castle Combe. All I can say is, I've been there too, and happily there's a long enough run-off so that if you keep going on the green part, you'll get back on to the black bit soon enough.

The Alfasud lends itself readily to conversion into a racing car, and there is a class specially for 'Suds and 33s in the AROC Championship series. Here at Cadwell Park is Ian Brookfield in his Ti, prepared at Spur Garage of Wimbledon, on his way to winning the 1997 Auto Italia Championship.

A typical paddock scene at Cadwell Park, Lincolnshire, with a selection of 'Suds, 33s and a GTV6, scheduled to compete in rounds of the AROC Championship in 1997.

The saloon car in the foreground is a 1983 Alfa 33. The racing car is also a 33, but it's a Tipo 33 sports prototype, with which Alfa Romeo won the World Championship for Makes in 1975. The racing T33s first appeared in 1967 with V8 engines and fibreglass bodies, and enjoyed some success. By 1971, when the big Porsche 917s were invincible, Autodelta ran three T33s for regular driver pairings of de Adamich/Pescarolo/Peterson, Stommelen/Galli, and Hezemans/Vaccarella/van Lennep. De Adamich gave Alfa its first major win in twenty years at Brands Hatch in 1971 at the BOAC 1000km. The 33TT12 seen here had a tubular chassis and used a 3-litre flat-twelve engine, clinching the title in 1975 with Ickx, Bell, Pescarolo, Andretti, Lafitte, Brambilla, Vaccarella and Merzario in the team, and was victorious again in 1977 when the 33SC12s had a clean sweep, taking all eight rounds. On the strength of this success, Alfa called the 'Sud's replacement the 33, although it's likely that not all customers were aware of the intended association.

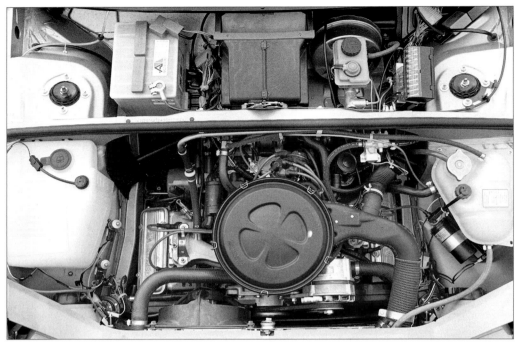

The 1983 Alfa 33 was powered by the same flat-four boxer engine as the Alfasud, and at first it came in two formats: 1.3- and 1.5-litres. Performance figures were not dissimilar, being capable of propelling the car to 100mph and 0–60mph in around 11 seconds. Pictured here is the single carburettor version with pancake air cleaner embossed with the cloverleaf symbol. The capacious windscreen washer fluid vessel sits on the right-hand inner wheel arch.

The Alfa 33 used the same platform and running gear as the outgoing Alfasud, but its bodyshell was more up-to-date, as well as bearing some family resemblance to the new Giulietta saloon. The stylistic treatment of its rear three-quarter panel also anticipated the angular uplift that would characterise the bigger Alfa 75 that appeared in 1985. This 1984 model wears the 'telephone dial' alloy wheels.

As you'd expect for a company that specialises in sporting vehicles, Alfa estate cars are rare, and manufactured by specialist coachbuilders like Boneschi, Pavesi, Zagato, Fissore and Colli. Harking back to the Alfasud Giardinetta was this 1.5-litre 33 Giardinetta shooting brake, which came with four-wheel-drive – using Subaru-sourced components. Unlike the 33 saloon, the shells were built by Pininfarina at the Grugliasco factory and then finished off at the 33's Pomigliano car plant. The Pininfarina badge adorns the Giardinetta's C-pillar, testifying that the extension to the regular saloon body was designed by them and not done in-house. My wife Laura had a 33 Sportwagon that succeeded the Giardinetta and was just front-wheel-drive, and it proved to be a rapid vehicle for making cross-country runs, being more agile in narrow lanes than a bigger car. As to its practicality, that was less convincing.

The 33 Giardinetta provided a fair amount of carrying capacity, especially with the rear seats folded down. The 33 Sportwagon was basically the revamped Giardinetta, and not having four-wheel-drive meant that the boot floor was lower, which improved its carrying capacity. The real handicap to this design was that the estate car hatch did not extend below the number plate, so the access to the cargo deck was not brilliant. For instance, you could get a bike in, but you could forget about a washing machine!

Just as some examples of the Giulia Nuova were fitted with a diesel engine, so it was with the 33, and indeed the Alfetta saloon. The 33 TD was a rather different proposition, however, producing a more sparkling 83bhp. The motor was sourced from Stabilimenti Meccanici VM, an associated company within the government-owned IRI group. It was a 1.8-litre three-cylinder unit, equipped with a turbocharger and intercooler.

The interior trim of the 1.7IE and TD 33 was plastic and upholstery was cloth-covered, which gave it a certain practicality, but it didn't exude quality and couldn't be described as outstanding. This is slightly more jazzy than the standard trim of the base model. For the taller person the seats and ergonomics of the cabin were best described as inadequate.

Viewed in profile, the 33 is a neat and inoffensive small saloon. This is a 1.8 TD of 1986, identified by the small visor that forms a canopy over the top of the rear window. Alfa tried a high performance diesel in 1982, fitted with a Comprex supercharger, which took it from 0–60mph in 12 seconds and gave a top speed of 108mph. Economy was good too, with up to 47mpg achieved in tests.

Here's an Alfa publicity photo of the six models in the 33 range in June 1988. The base model is the 1.3S, rising to the 1.5TI and 1.7 IE with the 1.7 Cloverleaf model with its boot spoiler at the sporting end. Odd-balls are the 1.5 4×4 and the 1.8TD with rear window visor. Absent from the line-up is the Arna, the result of a joint venture with Nissan, assembled at Alfa's Pratola Serra plant, and basically a Nissan Cherry Europe body with the 1.5 Alfa 33 engine installed and an Alfa grille on the front.

This is a racing car? You'd hardly credit it, looking at those anonymous Euro-bland lines. But the Arna – a discreet mixture of Nissan Cherry Europe and Alfasud mechanicals – won the UK's AROC Championship in 1997 and '98 by virtue of accumulated Class C wins in the hands of Dave Streather. There's no particular secret: once the Nissan N12's rear suspension set-up was sorted out, it was essentially the same as an Alfasud or 33 clad in a different shell. Confirmed *Alfisti* will no doubt try to draw a veil over Streather's successes, but there's no getting away from the fact that the Arna was assembled in Italy, and it does wear an Alfa badge. There were two- and four-door versions, and it came as a basic and rather sluggish 1.2 or a 1.5 GTi with bright green seats and a pair of Dell'Orto or Weber carbs.

By June 1988 the 33 was available with a fuel injected version of the boxer engine, with capacity rising to 1.7 litres, in which format it was fitted with hydraulic valve lifters. It proved to be Alfa's best-selling model, with 990,000 units built during its eleven-year history. That was approximately 80,000 more than the Alfasud, which remained in production for twelve years.

The 33 was a heavier car than the Alfasud and therefore less agile, but it was an adequate performer and fun to drive in a rural environment, coming into its own on fast A- and B-roads where its flat handling characteristics and good turn-in gave it an advantage. Acres of dark rubberised plastic trim marred the appearance of all Alfas of the 1980s, and only worked if the car's body colour was black.

This 1.7 Cloverleaf sports standard-issue alloy wheels and boot spoiler. Other subtle aerodynamic embellishments include the lip at the trailing edge of the rear valance, and the embryo side skirts on the sills integrate quite well with the wheel arches. The full Veloce body kit included a splitter below the front air dam, and the whole ensemble was also available on the 33 Sportwagon.

By the mid-1980s body kits were becoming commonplace, derived largely from the aerodynamic experiments of racing teams. Several Alfas received the race-replica treatment, including the 1.7 Cloverleaf of 1986, here showing off its front air-dam.

This punchy looking package is the 33 Permanent 4, which inherited its four-wheel-drive set-up from the Giardinetta, although the system was extended from part-time to permanent mode. In practical terms, its effect was to subdue the torque-steer wrestling match that had been the bane of the powerful 1.7-litre 33s, and it made the 33 more sure-footed than ever. Relocation of the suspension mounts contributed in part to this. Power came from the quad-cam 16-valve 1712cc flat-four motor introduced in 1990, which produced higher torque at lower revs and attained a useable 137bhp at 6500rpm, taking it to a potential top speed of 128mph. Economy was a not unreasonable 29mpg.

The interior of the 33 Permanent 4 featured Recaro-style front seats that provided excellent lateral support and an adjustable squab that extended forward to give better under-thigh support. The level of trim and cabin furniture like the door pulls and arm rests was also raised to a level commensurate with the car's other abilities. The instrument binnacle no longer moved up and down with the steering wheel adjustment as had been the case earlier, but the pedals were still too close together for anything over size nines, and the relationship of pedals and wheel to seat remained a problem for taller motorists. The downside of the rearward transmission was that the cargo deck was slightly higher, with consequent reduction of luggage space.

Racing at club level can be just as fast and furious as big league contests, and nowhere is it more competitive than in the AROC Championship's classes E and F, which are largely the province of 'Suds and 33s. Here Graham Heels leads Dave Ashford's similar 1.5-litre 33 at Brands Hatch in 1997.

Government pressure made Alfa Romeo build a new factory at Pomigliano d'Arco near the labour-rich city of Naples in the late 1960s. It was here that the Alfasud and 33 were built, and the southerly location – as opposed to Alfa Nord at Arese – gave the Alfasud its name. The entire project was managed by former Cisitalia engineer Dr Rudolf Hruska (who had also worked with Professor Ferdinand Porsche on the original project VW). Hruska had also overseen the launch of the Giulietta in 1954, and directed the engineering of the 'Sud. Here are the metal presses at Pomigliano, with appropriate sound insulation.

By the time the 33 came out the Pomigliano lines were fully robotised, which was rather at odds with the original thinking behind locating the plant where labour was plentiful and potentially cheap. Here we see the robot welders constructing the 33 inner shells.

Gone were the days of the bloke with the spray gun. Even the painting process was automated for the 33, as the shells were drawn along an aerial track like trolley buses. The pigment was applied by means of rollers from above and by rotating cups from the sides.

Quality control staff appear to be concentrating on the rear hatch of a newly finished 33 as it receives its check-over at the end of the line at Pomigliano. The wall chart indicates that the date is 11 May 1988, and that the diminutive Lancia Y10 is also assembled here. Once finished, Alfa 33s were put to the test on the Pomigliano proving ground, which included a high speed circuit with banked bends, a variety of different road surfaces and a hump-back bridge.

Chapter Five
Famous Names Return

When you've found a winning formula, there's every reason to capitalise on it, and this is what Alfa did when they needed to identify their new models in the 1970s, reviving the Alfetta and Giulietta names from its halcyon days.

To find a name for the 2000 Berlina's replacement, Alfa Romeo delved into the archives and came up with the Alfetta. This referred to the all-conquering Type 158 and 159 post-war Grand Prix cars that won 47 out of 54 Grands Prix they entered, taking Farina to the inaugural Formula 1 World Championship in 1950 and gaining Fangio the title the following year. The similarity with the 1970s saloon car was mechanical rather than sporting, and the justification for applying the Alfetta name lay in its transaxle design and de Dion suspension. This single headlight model is a 1.6-engined car, introduced in 1974.

The Alfetta was built at Arese and launched at a Trieste press show in 1972. It displayed more balanced, linear styling than its predecessor. It was powered by the familiar aluminium four-cylinder twin-cam engines, initially of 1.8 litres – the old 1779cc unit with different sump – fed by two twin-choke sidedraught Dell'Orto carbs, and when it reached the States in 1974 Spica fuel injection was fitted on US cars. The five-speed gearbox was mounted in-unit with the final drive, endowing the car with a well-balanced weight distribution virtually 50/50 front and rear.

Seen here at an AROC UK club meet, the Alfetta 1.8 was the first Alfa production car to use ZF type rack-and-pinion steering. Suspension was independent all round, with new and lighter components including longitudinal torsion bars at the front for the first time, upper and lower wishbones and caster rods, coil springs and dampers all round, anti-roll bars front and rear, plus the de Dion axle and Watts linkage at the rear, all conspiring to give more positive handling characteristics. The rear disc brakes were mounted inboard on the transaxle in order to further improve the unsprung weight.

The dashboard of the Alfetta was logically laid out, although the effect is rather cluttered, lacking the stylishness of the old Berlina. It was a more spacious interior than the 105-series models, however. This is an appropriate point to remind ourselves that production of the 105-series only ended in 1976, and as is the case with nearly all Alfa model changes, there was a fair degree of overlap between 105- and 116-series. However, the driving position of the Alfetta was still frustrating for the northern European physique, and the most harassing function of all was the gear shift, which was vague and imprecise. There was no such thing as a slick shift with an Alfetta, although it improved with later cars. The reason for its inexact nature was the length of the linkage that extended from cockpit to transaxle. Specialists improved the shift quality by lubricating the gear lever pivots.

The cabin of the 1972 Alfetta was comparable with that of most family-sized saloons, but ergonomics and special touches of trim like the door handles and window winders lifted it above the average. The relative size of the Alfetta meant that its seats and controls were bigger and therefore more comfortable than the Alfasud's.

The improved weight distribution derived from its rear transaxle gave the Alfetta more neutral handling characteristics than previous models and improved traction on a wet road. Housed in the transaxle assembly were the differential, transmission and a flywheel with a self-adjusting hydraulic clutch. A limited-slip diff was optional at first and became standard later on.

By 1977 the Alfetta range had been expanded to include the 2000, marked by a mild facelift and revised rubber-clad bumpers, intended to satisfy safety legislation, instead of the far more elegant chromed ones on the early cars. North American market cars – known as Alfetta Sport Sedans – were further burdened by mandatory heavy-duty 5mph-impact bumpers. The Alfa grille-shield was flattened out, and rectangular headlights further compromised the model's good looks.

The Alfetta 2000L of 1978 was available with three-speed ZF automatic transmission, which effectively negated any criticism of the manual shift: if you didn't like it, you could buy a fool-proof automatic. What could be better for urban use? However, once an Alfetta was up and running in the high ratios, the gearchange became less relevant and you could enjoy that almost indefinable planted quality that most Alfas have. The automatic Alfettas were unique in having a hydraulic self-levelling suspension facility that ran off an engine-driven pump. They also had a transmission fluid warning light and push button tester.

Alfettas were mainly pretty reliable, but there were problem areas that affected the whole Alfetta range, including the GTV coupé that came out in 1974. These included a tendency for wheel bearings to wear and head gaskets to blow (in my experience), and front wheel bearings became noisy if they were over-tightened, while the three rubber doughnut couplings in the driveshaft were prone to splitting. And as with all cars of this era, Alfettas could easily succumb to corrosion. The clips and screws that attached external trim broke the surrounding paint finish and rust took hold. Mud would build up on a ledge at the rear of the inner front wheel arches and eat away around the scuttle and base of the A-pillar. This is a twin-headlight fuel injected car with 1979-style alloy wheels. Late model SL Alfettas were finished in two-tone colour schemes and were considered to be very good value in the classic car marketplace.

The interior of the 1982 Alfetta had matured, and featured a wood-rim steering wheel and golf-club type gear lever knob. Switchgear was still located rather haphazardly, but the overall effect was better co-ordinated. The specification now included headlight wipers that were fed with a washer system.

Another great name from Alfa's past was resurrected for the Alfetta's sister car, called the Giulietta. Launched in 1977, it was another three-box saloon, tending towards a wedge-shape and with more angular styling and truncated tail than the Alfetta whose floorpan it was based on. Suspension and running gear were identical, but external features like the lip at the trailing edge of the boot, and the internal trim, were sufficiently different to cater for divergent market requirements. Its carrying capacity was obviously less than the Alfetta's, and this tended to suggest a more sporting character, which was not really the case as the Alfetta was just as much a driver's car.

During its production life-span the 116-series Giulietta was fitted with each one of the straight-four aluminium twin-cam engines, beginning in 1977 with the 1.3 and 1.6 units – the Alfetta never had the small capacity engine – and rising to 1.8 litres in 1979.

Top of the Giulietta range in 1979 was the Super, fitted with the 2-litre twin-cam. It was distinguished from the smaller engined models by the unattractive 'Super' graphic below the A-pillar and contemporary Campagnolo alloys.

The dashboard of the Giulietta 2.0 Super of 1982 housed the semi-circular speedo and rev-counter plus auxiliary dials and warning lights in a central binnacle ahead of the leather-rim wheel.

Just as the whole range of engines was available in the Giulietta, so the complete list of traditional nomenclature was applied. This is the 2.0 Ti of 1982. This front three-quarter shot demonstrates the compact wedge look that lingered on in sporting cars such as the Lotus Excel, TVR 350i and Maserati Biturbo. Headlamp wipers and driving lights were also fitted.

By 1984 the Giulietta 2.0 instrument binnacle had adopted a semi-circular housing, while the centre console was similarly rounded off to match. Switchgear was relocated to the centre and more thought given to storage bins.

It may look like any other Giulietta, but this is something a bit special. It's a 155bhp turbocharged model, prepared by the works' racing off-shoot Autodelta. Based on the Garrett T3 unit, it had originally been developed for the racing Alfetta GTV Turbodeltas, but was adapted for a limited run of GTV and Giulietta road cars. In the UK a number of GTVs were turbocharged by Surrey-based specialists Bell and Colvill in the late 1970s. Cosmetic differences on the Giulietta Turbo extended to driving lights integrated into a front apron and 'telephone-dial' alloy wheels.

Just as the 2000 and 2600 Berlinas had fulfilled a niche in the market for what we now call executive-sized cars, so the Alfa 6 was introduced in 1979 to perform a similar role. Although it was based on the Alfetta platform, the Alfa 6 had a longer wheelbase, and was the first Alfa to use the 60-degree V6 engine that has subsequently received so many plaudits. The 2.5-litre Alfa 6 didn't have the transaxle arrangement of the Alfetta and Giulietta, although it retained the de Dion back axle and inboard rear brakes. UK versions were supplied with ZF three-speed automatic transmission with hydromatic torque converter, while elsewhere it was available with five-speed ZF manual box.

The Alfa 6 was an imposing car, with a considerable amount of overhanging bodywork at front and rear. Nevertheless it remained a handsome car, in profile at least. Head of Alfa Romeo's design department at the time was Carlo Bianchi Anderloni, who had worked at Arese since 1966 when Touring Superleggera folded, and he expressed himself in the press as being quite content with the Alfa 6's styling.

In company with the Alfetta, the Alfa 6 was given rectangular headlights in 1982, together with a wash-wipe system and a splitter on the lower edge of the front valance. The carburettors of the original engine were replaced with Bosch Motronic fuel injection, which gave vastly improved fuel consumption.

Another variation on the Alfetta theme was the Alfa 90, which was brought out in 1984. Bertone was engaged to tweak the styling, and the 90 was powered by the 2.5-litre V6 engine. In this case the layout of the drivetrain was the same as the regular Alfetta, with a transaxle incorporating the clutch, gearbox and final drive.

Predictably, the dash and instrumentation of the Alfa 90 were little different from those of the Alfetta and Giulietta, dominated by rectangular panels and swathes of black and grey leather-look plastic. Heater controls were the same as in the new Alfa 75 that came out in 1985.

Alfa's model ranges can be viewed as a sequence of layers, with cars like the Alfa 6 and Alfa 90 overlying the Giulietta and 75, which in turn overlap the Alfasud and 33. Thus the well-proportioned Alfa 90 was the upmarket version of the Alfetta, although in retrospect it was something of a low-volume stop-gap.

Velour upholstery and trim on the seats of the Alfa 90 was both neat and practical. Rear seat legroom was just about adequate provided the front seats weren't pushed right back.

This is what all the fuss was about: the 2.5-litre V6 engine, installed here in the Alfa 90. Mounted fore–aft, the engine's Bosch L-Jetronic injection system with its prominent plenum chamber sits between the two banks of the 60-degree V, with the belts for the single overhead camshafts encased within the plastic cover.

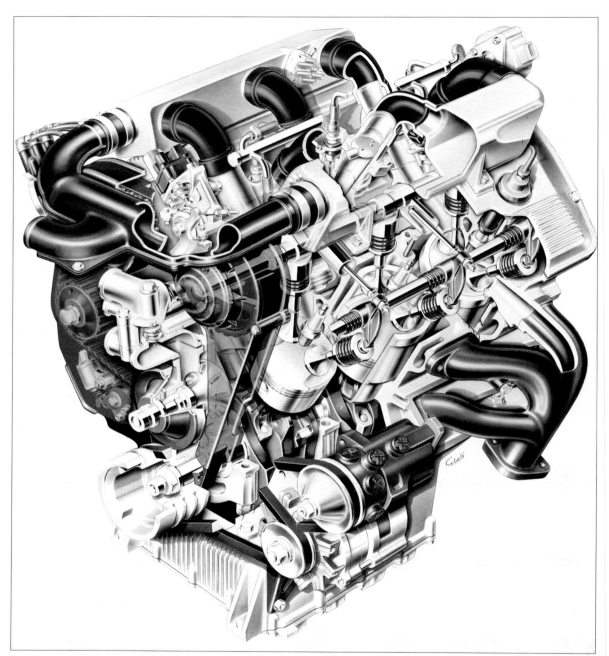

The anatomy of the V6 revealed. Construction is all aluminium, with 9.0:1 compression ratio and bore and stroke of 88 x 68.3 giving 2492cc. Maximum power is 156bhp at 5600rpm, and 154.9lb/ft torque at 4000rpm. What this doesn't tell you is that its most distinctive feature is the glorious bark that it makes, rising in a wailing crescendo as it hits high revs, making the Alfa V6 engine almost as seductive as the car it's powering.

This Alfetta is a Class E competitor in the AROC Championship in 1998. The netting on the window is to protect the driver from debris while allowing maximum ventilation.

An operator loads one of the sheet metal presses at the Arese plant in 1980 to fabricate a section of Alfetta bodywork.

Operators on the Alfa 6 production line install the V6 motor into the engine bay. Work had started on the V6 unit back in 1971 as an intended replacement for the familiar in-line twin-cam, under the direction of Edo Masoni in Alfa's engine development department. At first Masoni considered adapting the V8 used in the original T33 sports prototype and Montreal coupé, but this was rejected as being too complex. Thus the V6 was a brand new design, incorporating narrow-angle valves and a single hydraulically tensioned belt-driven camshaft per bank.

As was customary practice, Alfa Romeo produced a coupé model alongside its contemporary saloon, and in 1974 the Alfetta GT was announced. At first it was fitted with the 1779cc twin-cam four, and became known as the 1.8 GT with the arrival of the GTV 2000 and 1600 GT in 1976. Pictured here is the GTV 2000, its superb Giugiaro-designed body combining crisp linear styling with sensuous curves to provide hatchback practicality (apart from its central gas strut support) with genuine two-plus-two accommodation inside. I ran two of these cars, including a Strada model that had electric everything but corroded before your very eyes, and notwithstanding the notorious Alfetta gearshift, they were a real pleasure to drive. These chrome-bumper models featured a quirky instrument layout that placed the rev-counter in front of the driver and everything else in the centre of the dashboard and console.

In 1980 the Alfetta GTV underwent a thorough facelift which replaced the chrome bumpers with rubberised black plastic ones, which included a front air dam and minimal sill skirts. The overall effect was to give the GTV a more purposeful appearance, at the expense of some delicate detailing, which makes the early models perhaps more aesthetically appealing. Alongside the 2-litre car was the GTV6, also introduced in 1980, which was powered by the 2.5-litre V6 engine and race-prepared cars were campaigned with much success in touring car competitions the world over. It is still much in evidence in the AROC Championship, and my own car is seen here rounding the Mallory Park hairpin in 1990 as I struggle to keep ahead of an Alfasud.

Alfa Romeo's more recent presence in Formula 1 was not marked by any enduring success. Here are Andrea de Cesaris (left) and Mauro Baldi with the 183T turbocharged V8s in the paddock at Long Beach in 1983 for the USA West Grand Prix. They were by no means hopeless, because although de Cesaris' gearbox broke and an accident put Baldi out of this event, de Cesaris finished second in both the German and South African Grands Prix and fourth at Brands Hatch, while Baldi was fifth at Zandvoort and sixth at Monaco. De Cesaris was eighth in the final points standings while Baldi was a lowly sixteenth, in a year that saw Nelson Piquet crowned champion and Ferrari take the constructors' prize.

In 1985 Alfa Romeo celebrated seventy-five years of car manufacturing, and marked the event with the eponymous Alfa 75. Financial dictates meant that it was not possible to design an entirely new model, so the 75 was something of an amalgam of Alfetta and Giulietta styling and running gear, albeit with a number of improvements. The 75's suspension system was similar to that of its predecessors, and it too used the transaxle arrangement for gearbox and clutch. First model off the Arese production line was the 1.8, which didn't run to a limited slip differential.

The 75 looked better in a dark colour scheme, which disguised most of the quirky styling cues like the heavy bumpers and the angled plastic rubbing strip on the rear three-quarter panel. Its compact proportions belied the fact that it had quite a capacious boot, which extended underneath the rear window to provide 17.5 cubic feet of space.

The curious design features weren't confined to the external styling. Within the cabin, the window buttons that activated the 75's front windows were located in the centre above the windscreen, along with the interior light and map-reading light. Another peculiarity was the handbrake lever, which was shaped like the handle of a spade. Once you got used to these idiosyncrasies though, they seemed quite normal. My mother-in-law had an early 1.8, which had a little row of vertical yellow lights on the edge of the dashboard, with an arrow to indicate when you ought to change gear.

The 75 was marketed as the Milano in the USA.

In 1987 Alfa's venerable twin-cam received a shot in the arm with the productionising of the twin-spark system, hitherto the province of the GTA racing engine although Alfa employed two spark plugs per cylinder as far back as 1923. The advantage was that it provided more efficient combustion without the complexity of multi valve heads. The 75 was equipped with the 2-litre Twin-Spark engine in 1987, allied to computerised Bosch Motronic engine management that controlled ignition and fuel injection. Note the second distributor mounted on the front of the engine ahead of the oil filler cap. Its smooth power delivery made the 75 reliable, economical and a class leader in the performance stakes. The Twin-Spark engine was also used with much success in Dallara and Reynard Formula 3 single-seaters by the likes of Bertrand Gachot, Thomas Danielson, Phillipe Favre and the late Roland Ratzenburger.

As well as the 1.8 and 2-litre TwinSpark models, the early 75 was equipped with the 2.5-litre V6 that previously graced the Alfa 90 and Alfa 6. The 2.5 V6 was also available with automatic transmission, and its output was 156bhp at 5600rpm.

In 1989 the Tipler stable included a 75 TwinSpark and GTV6. One was turned over, almost into oblivion, on the back road between Chepstow and Tintern, and the other was turned into a racing car. The 75 was formerly an Alfa GB press car, fitted with Veloce body panels and the more useful addition of a Harvey-Bailey handling kit, which consisted of stiffer springs and thicker anti-roll bar. Its supplier was Derbyshire-based suspension guru Rhoddy Harvey-Bailey, a former Autodelta driver who drove the entire Snetterton 500km single-handed in 1968 in his GTA.

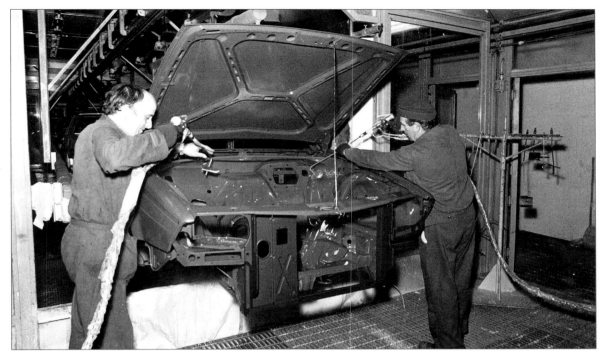

In a bid to counter the Alfetta's poor reputation for corrosion, the 75 received special attention in preventative treatment. Here in the Arese bodyshop, operators inject wax into cavities in the scuttle that might be prone to rust. In fact, the problem areas on the 75 tend to focus on the rear light clusters, but only become apparent after ten years or so.

On the Arese production line, 75 bodyshells with their interiors installed have sections of exterior trim fitted. The seats were upholstered in durable material, flecked herringbone velour in the top-of-the-range models and coarser cloth on the 1.8. The seams on a much-used rear seat squab were prone to split where cloth met vinyl, however.

It looks like any other regular 75, but the car on the left in this Normandy hotel car park is powered by a turbodiesel. British motorists have never had to face the temptation of the 2.4-litre VM-engined cars, but in continental Europe, where oil-burners enjoy a running cost advantage, they do quite well.

The badging proclaims this 75 is propelled by the turbocharged 1.8-litre twin-cam, which, like the diesel, was never available in the UK. The installation was based on the ubiquitous water-cooled Garrett T3, although racing versions used the KKK turbo, and included an intercooler and fuel injection. The 75 Turbo was developed into a competitive touring car racer under the direction of Alfa Corse's competitions manager Giorgio Pianta, himself a GTA racer of no mean ability in the 1960s and former Lancia rally chief.

This is what the factory-built and works-supported 75 Turbos and Evoluzione 3-litre racers looked like in race trim, with full front air dams, side and rear skirts, and boot-mounted aerofoil. The exhaust pipe emerges through the side skirt under the passenger door, while inside there is a racing seat and full roll-over cage that had the effect of stiffening the shell. Drivers in the Alfa Corse squad included Jacques Lafitte, Jean-Louis Schlesser, Michael Andretti, Nicola Larini, Thomas Lindstrom, Gabriele Tarquini, Rinaldo Drovandi, Paulo Barilla and Giorgio Francia.

One of the most successful drivers in the AROC Championship series in the late 1990s was Graham Presley, whose 75 was powered by the 1.8-litre turbo engine. The car graced the national AROC meeting at Stanford Hall in 1997 to showcase the talents of its builders BLS Automotive of Lincoln. A year on, it was completely re-shelled following a major accident, but undaunted Presley just carried on winning.

The competition image of the Group A 75 racer was applied to the road-going 75s, which were designated Veloce models, resplendent in the race-replica body kit and Revolution alloy wheels. The 3-litre model was first imported into the UK in September 1987, and this is a 3-litre V6 Veloce that I ran for over 100,000 miles, discovering in the process that its plastic protuberances proved vulnerable against kerbs, and were mounted so inefficiently that the slightest nudge could displace the whole frontal ensemble from the bumper down. I cannot comment on their aerodynamic effectiveness, but it was a thrilling car and much underrated, with bags of power and decent handling improved further by lowering the suspension, and fitting aftermarket dampers and Yokohama tyres. Shame about the brakes.

This gem is the 3-litre V6 unit, showing off no fewer than nine pulleys driven by belts for everything from cams and power steering to air conditioning. The bane of the 75's exhaust system was that the downpipes passed underneath the car and struck every traffic-calming hump and sleeping policeman, so that premature replacement was inevitable.

In 1990 the AROC Championship trophy was carried off by Roger Kay in his Alfa 75 3.0, pictured here at David's hairpin, Brands Hatch, sweeping all – including me – before him. He notched up twelve outright wins and set ten lap records. The car was prepared by York-based Alfa dealers Alwyn Kershaw, who had done a similar job on Kay's GTV6 the previous year. Apart from its well set up suspension, much of the car's excellence stemmed from its blueprinted V6 engine.

The factory never made a coupé to match the 75 saloon, but a limited run of a thousand two-seater supercars, based loosely on the 75, was created by Zagato in 1990. The ES30 SZ was dubbed **il môstro** – the monster – because of its bluff exterior, and it was built up on the 75 floorpan and running gear but clad in the most outrageous coupé body since the BAT cars of the 1950s. The regular 75's torsion bar front suspension was banished in favour of Koni dampers and coil springs, and rose joints replaced the more compliant rubber bushes throughout.

The powerhouse of the SZ was the 3-litre V6 unit, taken straight off the 75 production line and tweaked to produce 210bhp, an increase of 18bhp on the standard engine, with 184lb/ft of torque at 4000rpm. In such a lightweight shell, its outright performance was in excess of the 137mph credited to the standard 3-litre 75.

The cockpit of the SZ, with a pair of leather upholstered bucket seats, sculpted steering wheel and wraparound dashboard. If any Alfa was ever a supercar, this was it. This impression gained credence by the fact that the ride height could also be adjusted by 6cm to ground-hugging level by a pair of switches on the centre console, and its wide Pirelli P Zero tyres were compounded specially for the SZ.

Chapter Six

The Modern World

When Fiat took control in 1987 it was a time of transition. The 164 ushered in a new generation of front-wheel-drive cars in 1988, culminating in the 156 and 166 of 1997/8, placing Alfa Romeo back at the top of the saloon car manufacturing sector. The generation of cars that brought the company back into the limelight once more was powered by a new breed of engines, with some 'avant-styling' to match their performance.

The Pininfarina-styled 164 made its debut at the Frankfurt Show in 1987. It was considerably bigger than the 75, taking Alfa Romeo into a straight fight with cars in the so-called executive segment, like the BMW 5-series. For the first time in a couple of decades the scribes in the motoring press were heaping plaudits on an Alfa, because it was more of a mainstream car and had none of the styling and ergonomic glitches that they felt had marred its predecessors.

Some like it hot! The 164 in this celebrated Cannes location is a 1991 2-litre V6 Turbo, a model produced expressly to duck under the Italian tax break yet still provide cracking performance. The 164's origins go back to an agreement made in 1982 between Alfa Romeo, Fiat, Lancia and Saab to build a series of cars based on the same floorpan. These became known as Type 4 cars, and evolved into the 164, Croma, Thema and 9000 models. One cost-effective element in the strategy was the use of a front-drive engine and transmission package for each model. For Alfa Romeo this meant redesigning the engine range for transverse installation, and it also implied that the 75 was the last rear-wheel-drive Alfa to be made, which was a matter of some regret to driving enthusiasts as front-wheel-drive cars lack the fluency of rear-wheel drive.

The 164 was launched on to the British market in October 1988, and at first only the V6 in standard and Lusso trim was available. Four-speed ZF automatic transmission could be specified, and the following year the 2.0 Twin-Spark (seen here) arrived in the UK. Suspension was independent all round, with MacPherson struts at the front and anti-roll bars at both ends. Naturally, steering was power assisted, with ABS brakes acting on outboard discs.

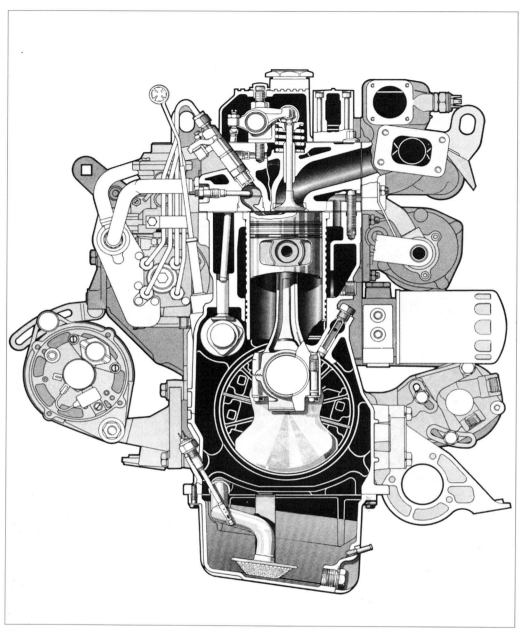

Continental buyers could opt for the 164 fitted with the VM 2.4-litre Turbodiesel engine, whose innards are revealed in this cutaway diagram. Since one of the inherent failings of powerful front-wheel-drive cars is torque steer, the TD is probably the least likely 164 model to fight with its driver under fierce acceleration. Alfa acknowledged that there was indeed a problem with torque steer on early 164s, and in 1990 brought out the Cloverleaf model with the V6 engine lowered in the engine bay, allowing the drive shafts to sit more horizontally. The front suspension was revised to include an electronically controlled semi-reactive variable damping system, and you could select either sport mode for a hard setting, or auto, and have the suspension automatically sorted for you according to prevailing road conditions.

Heavy investment in new manufacturing techniques – CAD and CAM – and the extensive use of robots and off-line assembly of key components ensured that the 164 was better made than any previous Alfa. Excellent build quality was tangible in the heavy clunk of the shutting doors. Anti-corrosion treatments extended to galvanising at least 60 per cent of the bodyshell, a PVC coating underneath, and wax injection of vulnerable box sections. Inside, the seats were fully adjustable electronically in most directions, and it was reasonably spacious in the rear for all except the tallest passengers.

This brute is the Alfa 164 Procar, powered by a normally aspirated V10 racing engine and built by Brabham to contest the still-born Silhouette series mooted by FISA to run in 1989. The idea was to have out-and-out racing cars masquerading as their regular road-going counterparts and clad in lightweight replicas of the specified model. This was not so far removed from the Class 1 International Touring Car Championship instigated by the FIA in 1996, except that in this 164 Procar the V10 is mounted amidships, whereas in the ITC the engine had to be located in its original position.

The Alfa 164 Procar was actually badged as a Twin-Spark. Who were they trying to kid? Its V10 engine had only a single spark plug per cylinder, and turned out to be a one-off. The de-mountable bodyshell was a convincing replica of the real thing, in Cloverleaf form at any rate, with its characteristic narrow band of rear lights and arrowhead nose that heralded the styling cue of a whole new generation of Alfas. The rear aerofoil should have been a bit of a give-away though, and those huge slicks half hidden under the wheel arches ought to have told a tale. As it was, the 164 Procar performed little more than shakedown tests in the hands of Riccardo Patrese. The Alfa Corse-built 3.5-litre, 72-degree V10 was potentially a Formula 1 engine and might have found its way into a Brabham, but nothing came of that either.

Would you let this man design your car? It would be a pity not to, since he was largely responsible for the current good-lookers from the Alfa stable. He's Walter da Silva, prime mover at Centro Stile based within Fiat's Turin headquarters. Centro Stile was set up in 1958 as Fiat's Central Styling Centre and Design School by Mario Boano and his son Paulo, where early work included the Fiat 850 Spider and Coupé and the Fiat 124 Coupé.

The Alfa 155 was introduced at Barcelona in January 1992 as a replacement for the 75. Although it was similar in size, it differed radically in being front-wheel-drive like the 164. Its proportions remained closer to those of the 75, but its styling was closer to the 164's. Although build quality was far better than the 75's, there were still some extraordinarily inconsistent panel gaps on the new model. A similar range of engines was available – 1.8 twin-cam, 2-litre TwinSpark, and 2.5-litre V6 (but no 3-litre) – and there was a four-wheel-drive version called the Q4. The car pictured is my 155 V6 Cloverleaf of 1995, which looked virtually identical to the Q4, including the 6.5J Speedline quasi-split-rim wheels, but differing in the flared wheel arches of the later facelifted model.

While the 155 V6 used the 166bhp 12-valve 2492cc engine found in previous Alfas – mounted transversely now – the Q4 used an Alfa-badged version of the 190bhp 16-valve turbocharged motor from the Lancia Delta Integrale. This made it the quickest of the 155s, but it had a reputation among the fraternity of Alfa specialists of being overly complex and difficult to fix. Available only in left-hand-drive, the Q4 was dropped in 1995. In practical terms, the regular front-drive four-cylinder cars were a bit nimbler than the V6, but there was no substitute for the sound and thrust of the six-cylinder car on full song. Torque steer was intrusive on a damp road surface though.

The 155 was based on the Fiat Tempra platform, along with the Lancia Dedra, and production lasted until 1997. Later models were facelifted with revised grilles and flared arches and are known as 'wide-body' cars. Significant handling improvements were also made in 1995, with wider track and lowered suspension, so the 155 was a well-balanced car with excellent turn-in. The V6 Cloverleaf of 1995 was unique in having an uprated suspension package as standard. There are hints of its Super Touring potential in the embryonic splitters that protrude from the front and rear skirts. The 155 is a more practical car than its predecessor in having a large boot, with bumper-level access for loading and a space-saver spare wheel stowed below the boot floor.

The driving position of the 155 was also better than in the 75, yet it was still not quite right for the northern European. The driver's seat on the V6 Cloverleaf had a facility for adjusting lumbar support, but the driving position was curiously upright by comparison with Alfa's later models. Upholstery was more resilient, and the transverse engine mounting and lack of transmission tunnel made for a more spacious cabin with slightly more legroom in the rear. Instrumentation was thorough and controls of better quality than before, especially the contoured Momo wheel. Principal dials were flanked by oil pressure and temperature indicators, with no fewer than twenty-two warning lights in a panel on the centre console below the heater and ventilation controls. Heated rear window and rear fog lights were fingertip sensitive buttons on the ends of lights and wiper stalks. Driving light switches were also on the centre console along with rear window switches.

The V6 motor was repackaged to fit the 155 engine bay under Alfa's head of engine design, Alessandro Piccone, while the TwinSpark unit was almost completely revamped. The V6 delivered power and torque smoothly and was mounted transversely in the 155, its chromed inlet manifolding a work of art. Engine management was a fully mapped Bosch M1.7 system, and a totally new compact alternator and micrometric tensioning system for the drive belt was also fitted. Shifting the gears of the five-speed manual box was by Bowden cables on the V6 and rods on the TwinSpark, and light years more positive than the transaxle arrangement of the old Alfetta-derived system, although it paid not to hurry second gear until the engine had warmed up.

Creature comforts in the 155 V6 – snapped here at Lanercost Priory, Cumbria – include power steering that decreases the faster you go, and a six-speaker radio cassette system. The later 155s had the so-called quick rack ushered in with the new Spider and GTV, requiring only 2.1 turns from lock to lock, although failure to maintain the power steering drive belt resulted in a screeching noise on full lock. The 155's Bosch ABS brakes were certainly a revelation after the uncertainty of the 75's less than satisfactory retarding properties. It had a diagonally split brake circuit, and a proportioning valve on the rear axle ensured that the braking effect between front and rear took into account the weight and distribution of the car's payload.

Sparks fly on the Arese production lines as robot welders piece together the sections that make up the bodyshell of the Alfa 155. Over 70 per cent of the 155's panels were galvanised, and the roofs on these cars already have sunroof apertures cut. Prototypes underwent some four million miles of test driving during pre-production development.

The original 155-based competition car was developed early in 1992 as the GTA and was based on the Q4. Under the supervision of Alfa Corse managing director Giorgio Pianta, general manager Pierguido Castelli and design engineer Sergio Limone, a front-wheel-drive version was built up to comply with the FIA's Class 2 regulations and tackle the burgeoning Super Touring category that was developing into a series of national championships all over the world.

They had a budget of £5m to contest the BTCC, and the 23-strong squad was managed by former Fiat Abarth and Lancia rally star Nini Russo. Driver pairings for the 1994 BRSCC were ex-F1 star Gabriele Tarquini and Giampiero Simoni. Four 155s were built up – two to race while another pair was refettled in Turin.

Anatomy of the racing 155, revealing the massive Brembo eight-piston calliper ventilated 15-inch discs at the front and four-piston callipers at the rear, with carbon metallic pads. Gas pressurised Bilstein dampers complement Eibach springs and MacPherson struts at the front, with fabricated steel trailing arms at the rear. Bolt adjusters on top of the suspension turrets tune the camber settings. The 155's hydraulic power steering was sourced from the Lancia Delta S4 Group B rally car. The shell is seam-welded and stiffened massively by the complex trusses of the roll cage. Wheels were 18in diameter × 8.25in wide Speedline-cast MIM using Michelin covers.

The Alfa Corse badge on the cam cover indicates that this is a very special engine. Working within the regulations of the organisers TOCA, Abarth built up the 155 Super Touring engine using components from a variety of Fiat corporate sources. The cast-iron block came from the 164 VM Turbo, the alloy head from the 155Q4 – basically an Integrale 16-valve – and rotated through 180 degrees to improve engine breathing. Power was transmitted via an AP racing clutch, solid steel driveshafts and a limited slip differential. The engine's internal dimensions were reworked and components sourced from a number of Fiat parts bins, and the whole unit was canted backwards at an angle of 27 degrees to compensate for the stresses and weight transference inherent in the front-wheel drivetrain and racing suspension. Peak torque arrived at 7000rpm, with 290bhp available at the mandatory 8500rpm limit. Abarth held a stock of forty engines for 155s active in various Super Touring championships.

When the 155s walked away with the victory on their first appearance in the 1994 BTCC there were howls of protest about their aerodynamic additions. These consisted of a two-position front splitter and big rear boot spoiler, evolved in the Fiat wind tunnel in order to eliminate lift and generate downforce. To appease the protestors the front splitter was retracted and rear wing extensions omitted, but Tarquini still took the Championship title. Simoni, seen here at Silverstone, finished the 1994 season in fifth place.

The cockpit of the 155 Super Touring cars was a combination of austerity, structural rigidity and hi-tech laboratory. The bare metal was criss-crossed with the bars of the roll cage, and characterised by the digital rev-counter, minimal switchgear and on-board computer, the sequential shift for the Hewland six-speed gearbox, suede-rimmed Momo wheel, Sparco seat and plumbed-in fire extinguisher system.

The 155's mechanical specification extended to a TAG 3.8 multi-point sequential and programmable fuel injection system with digital electronic ignition. The single 70-litre fuel cell was housed below floor level in the boot, filled through a carbon fibre funnel.

Although the 155 continued to be a major force in Super Touring elsewhere in Europe, the other BTCC teams had caught up by 1995 and the Prodrive-run team floundered with the unsorted suspension set-up of the new season's wide-bodied cars. Despite the best efforts of Prodrive's Dave Benbow, the 1995 driver pairing of Simoni (right), and F1 ace and World Sportscar champion Derek Warwick were struggling. In mid-season Giampiero was sent off to race in the Spanish Superturismo series, and Tarquini was brought back into the BTCC squad in a vain bid to regain lost ground.

Derek Warwick leads the 1995 BTCC pack through the tight Thruxton chicane, with a BMW M3 riding the kerbs, followed by Simoni's 155, with a Carina, an Accord and a Cavalier in hot pursuit. The Alfas may have been off the pace in the 1995 BTCC, but elsewhere it was a different story. For instance, Luis Villamil was Spanish Superturismo champion in a 155 TS.

Gabriele Tarquini's 155, with which he scooped the honours in the 1994 BTCC, contains a single Sparco race seat and six-point harness. A racing seat can be improved upon, and made to conform exactly to the driver's rear proportions. This is achieved by filling a bin-liner with two-pack insulation foam, which is placed in the seat, and as the solution rapidly expands, the driver sits in it and within seconds it has hardened to the appropriate shape, taking up any surplus space in the seat squab. The driver is thus held completely secure and can resist the powerful sideways forces generated under hard cornering.

In 1995 the FIA announced the International Touring Car Championship for Class 1 cars, a product of the high-tech German DTM series. Only three manufacturers participated – Opel with the Cosworth-developed V6 Calibra, Mercedes-Benz with their big AMG-built C-class model, and Alfa Romeo, using the 155 V6 Ti built by Alfa Corse. Maximum permitted capacity was 2.5 litres and six cylinders, and both Opel and Alfa were four-wheel-drive. They were comparable designs, which produced close and highly competitive racing from twenty-car grids.

The Class 1 Touring Cars that contested the ITC were out-and-out racing cars with special space frame chassis beneath the lightweight bodyshells. As is apparent from the 155, the V6 engine is low-slung in the engine bay to obtain as low a centre of gravity as possible. No turbos were allowed, but they still put out around 500bhp. The four-wheel-drive Alfa was equipped with an automatic computerised six-speed gearbox, with three electronically controlled diff-locks. The 155 had double wishbone suspension all round, hydraulically adjustable anti-roll bars that were controlled from within the cockpit, ABS brakes with eight-pot callipers, and power steering, with slick tyres on OZ racing wheels. One downside of the regulations was that top five success earned the driver a weight penalty, with up to 50kg (112lb) of ballast added.

ITC drivers came from a variety of backgrounds, including F1 stars like Christian Danner, Stefano Modena, Michele Alboreto, Sandro Nanini, Giancarlo Fisichella and Nicola Larini (among others) in the four teams running Alfa 155 V6 Tis. The ITC races were staged at a variety of venues ranging from converted street circuits like Helsinki to purpose-built autodromes like Hockenheim, and on tight tracks like Avus contact was inevitably frequent and violent. The circus came to the UK once for a round at Donington that was won by Tarquini, but despite much promise the 26-round series collapsed in financial and political acrimony at the end of 1996.

In 1994 Alfa Romeo came out with the 145, a three-door model that resurrected the old Alfasud theme. Its radical styling featured the arrowhead bonnet – or 'beak', as one classic car scribe put it – and several other features that made it stand out from the hatchback herd, including what they call poly-elliptical halogen headlights. The original 145 range was powered by the flat-four engine, so the comparison with the 1970s icon was even more apt.

Top of the 145 range was the Cloverleaf model, identified by the logo on the hatch and rear end of the side skirts. In original form it used the same 2-litre TwinSpark drivetrain found in the 155 TS. Stylistically, the 145 was as entertaining to look at as it was to drive – a relatively big engine in a small car is always a heady potion, especially if it's set up right. The rear three-quarter window of the 145 ensures good rearward visibility, while the shallow angle 'V' of the rear window matches the rear light clusters. The other amusing design cue is the dip halfway along the top edge of the doors.

Contemporary with the 145 was its sister car, the 146. This five-door hatch represented Alfa's compact saloon and could logically be seen as a successor to the 33. There is a wedge-like tendency about its styling with a swage line that rises along the flank to end in a high Kamm tail. Unlike on the 33, the rear window slopes down almost to the rim of the back panel.

The 1.5 and 1.7 boxer units that powered the original 145 and 146 were replaced two years on by the new generation of the evergreen in-line twin-cam four-cylinder engine. These state-of-the-art engines were developed thanks to Fiat rationalisation, and featured a cast-iron block that was reckoned to be stronger than the old all-aluminium unit. The new engine incorporated two spark plugs and four valves per cylinder, variable valve timing, hydraulic tappets, the latest generation Bosch Motronic engine management system, a steel crankshaft with eight counterweights, and counter-rotating balancer shafts for more refined running.

Interior of the 145 has comfortable Recaro-style seats with fold-down backrests and a sliding mechanism on the passenger's seat to facilitate entry into the rear of the car. The back seat folds down to create additional carrying capacity. The instrument binnacle ahead of the driver contains speedo, tachometer and water temperature and fuel gauges, while the centre console is angled towards the driver. The Momo wheel contains an airbag. The dashboard panel containing the glovebox curves sharply away from the front passenger seat to provide more legroom and enhance the feeling of space.

This cutaway rendering of the 145 reveals side impact beams within the doors, ABS brakes with ventilated discs at the front, independent suspension by means of MacPherson struts at the front, plus cast-iron wishbones, mounted with differential stiffness bushes, negative offset kingpins and stabiliser bars, with coil springs and dampers all round. There's an anti-roll bar at the rear, with longitudinal trailing arms mounted to the body via a cushioned subframe.

By 1998 the 146 was available in three forms, with the 1.6 Junior as the entry level model, the 1.8 T. Spark, and the range-topping 2-litre ti pictured here. The Junior was marketed as a descendant of the 1600 GT Junior coupé of 1974, and featured its own logo, distinctive 15in alloy wheels and colour-coded side skirts, door handles and rear wing. Although perfectly well appointed, the 1.8 T. Spark was the plain Jane of the range, while the 2.0 ti offered everything that the other two did, but with the obvious gain in performance.

Driver's eye view of the 146 ti dashboard, with three-dial instrument panel visible through the three-spoke leather-rim wheel. The wheel boss contains the driver's side air bag. Heater controls and vents dominate the centre console to the left, which also contains the stereo with its anti-theft pop-off front. Electric window switches and door mirror adjusters are housed in the inner door panel. Left-hand-drive cars were given a revised fascia in 1998 but the 145 and 146 were sold in insufficient quantities to make it worthwhile producing the new dash in right-hand-drive format.

Opposite: At the heart of the mid-range models in the 145 and 146 ranges was the 1.8-litre 16-valve TwinSpark engine. The light-alloy 16-valve twin-plug cylinderhead was unique to Alfa Romeo, and the steel block had higher tensile strength than the aluminium used previously. The TwinSpark engine required no attention – other than oil changes – between 100,000km servicing. It belonged to the modular family of Alfa engines built at the company's new Pratola Serra plant at Avellino, south of Naples, which came on stream in 1997. The aim there was to exploit modular construction in the interests of a more streamlined and efficient production process. Engine capacities ranged from 1.4- through 1.6- and 1.8- to 2-litres. Rather ironically, this factory was built on the site of the old Arna plant, scene of the short-lived joint venture with Nissan in the mid-1980s to hybridise the Cherry Europe with the flat-four motor of the 33.

Now is this not the most fetching of any car produced in the 1990s? Few would disagree that it has to be the best-looking saloon that Alfa has ever made as well. The 156 was launched in 1997 to rapturous acclaim, and its fabulous styling dispelled any doubt that the company was not on the right track. By this time most people were beginning to take the performance and handling aspect of modern Alfas for granted but, true to form, the 156 did not disappoint there either. There were engine options to suit most requirements, with the 1.8- and 2-litre TwinSpark models and the 2.5-litre V6.

Opposite, top: If the specification of the 156 was thoroughly modern, some of the styling cues evident in its gorgeous curving bodyshell were pointedly retrospective. The front door handles take you back to the Giulia Super, while its very roundness recalls the little Zagato Giulietta SZ of 1959.

Opposite, bottom: There's an ambiguous touch to the 156 bodyshell too. By means of concealing the rear door handles in the black trim of the C-pillar, it has been made to look like a two-door coupé from oblique angles. The panel fit is so good that the door shut lines are barely discernible. In keeping with other vehicles in the Fiat group – and its Alfa siblings – the light clusters front and rear have that compressed, minimalist aspect. Bumpers are reduced to thin rubbing strips at the apex of the bulbous valances.

Above: The top-of-the-range 2.5-litre V6 engine was endowed with four valves per cylinder, which lifted power output to 190bhp – that's close to the old 12-valve 3-litre unit's. Impulses from the electronic fly-by-wire throttle are transmitted directly by a control unit built into the engine, which regulates the rate of fuel intake by constantly monitoring the engine rpm, speed and gear ratio in use. The V6 model drove through a six-speed gearbox, which some might consider a novelty rather than a necessity. Nice to have the extra ratio to play with, though.

Opposite, top: Everything about the 156 seems so positive that it seems there must be a catch. The one downside becomes apparent once you've sat in it. And that is, how on earth are you going to get your hands on one? Yes, this sad eulogy extends to the interior, in which the seats are comfortable, the driving position excellent – at last – and the controls equally complementary. The principal dials are housed in twin nacelles that remind you of the 1750 Berlina, while the auxiliary gauges are angled towards the driver. The use of wood veneer on the centre console of the two bigger-engined models was Alfa's single mistake – I suppose it does complement the wood-rim wheel – but you could opt for the imitation titanium trim, which together with the leather rim wheel is standard on the 1.8.

Opposite, bottom: So as not to intrude on the classic Alfa shield-shaped grille, the numberplate was positioned over to the left of the car, which is where classic Spider owners were accustomed to putting it. The 75's Veloce body styling and the 155's Silverstone limited edition provided precedents for the Sport Pack cost-optional accessories that 156 owners could use to personalise their cars. These ranged from low-profile tyres with lowered suspension and Superturismo-style wheels to a gargantuan rear wing and side skirts and splitters. To gild the lily yet further, Momo-designed leather seats were optional.

The door apertures on the 156 are wide enough for easy access to the cabin, but the boot opening might be considered less ample when stowing larger items of luggage. Carrying capacity would be quite adequate for most occasions, though.

The 156 rear light clusters extend inwards from the outer wing panels to cut into the boot lid, and are a progression of the design features on the 145 and 146 models.

You can't see them any more in the redesigned engine casing, but each combustion chamber in the Twin-Spark motor has two platinum spark plugs, of different sizes and performing different roles. The larger one is central, and fires the compressed charge at the beginning of the power phase, while the smaller one is offset at one end of the chamber and sparks 360-degrees later at the end of the exhaust phase. This has the effect of both reducing emissions and protecting the catalytic converter by ensuring that no unburned fuel can reach it. Each cylinder has its own direct-ignition coil, and the spark plug leads are arranged so the output from each coil is directed to two different cylinders. Because four-cylinder engines are inherently less well-balanced than sixes, Alfa Romeo builds two contra-rotating shafts into the TwinSpark motor to iron out the alternating moving masses which cause the imbalance. The crankshaft carries eight counterweights and a torsional vibration damper, and a cast-aluminium alloy sump adds stiffness to the engine and gearbox assembly. The 156 is also made with the five-cylinder 2.4 JTD turbodiesel engine that uses the fuel-efficient Unijet common-rail direct injection system.

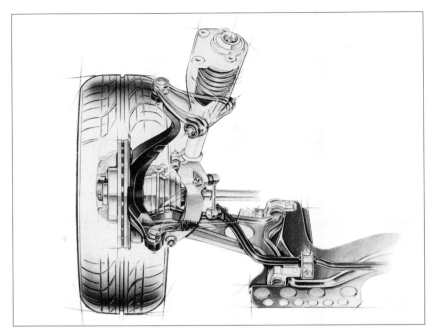

Alfa saloons get more and more sophisticated. The front suspension of the 156 consists of double wishbones, coil springs and dampers, with rose-jointed anti-roll bar connected to telescopic struts. This layout is designed to ensure an anti-dive effect so that the front end doesn't dip under braking or lift under acceleration.

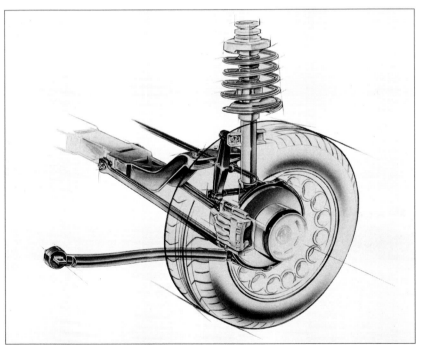

The 156 rear suspension is a departure from traditional practice, comprising MacPherson struts with differential length tie rods connected to an aluminium beam, with longitudinal reaction rods, offset coil springs and dampers. The set-up is completed by a rose-jointed anti-roll bar, and the overall aim is to control those undesirable but inevitable motions of pitch and yaw, and create a self-steering effect that enhances the car's stability when driven hard.

I was lent a 2.5 V6 model by my local Alfa dealership, and was impressed by its free-revving engine and slick gear change. Putting it through its paces, acceleration was suitably swift, and despite the belief that the V6 model is less sprightly than the 2-litre car, I found it was still pretty agile.

Turn-in was sharp and you could clip any apex you cared to aim at with total accuracy. There appeared to be little tendency to understeer, and it was a pleasure to drive quickly. As seems to be the case with the current generation of Alfas, there is a pronounced thud as you drive over serious undulations like traffic-calming humps and dips or potholes. As my 155 displays similar characteristics, I assume this is down to harder suspension settings and wider, low-profile tyres.

One of the colours in Alfa's paint-shop palette is called Nuvola Blue. Depending on the prevailing light conditions when you see a car in this pearlescent hue it can appear silver, pastel blue, even slightly pink or gold. They charge you extra for this variable paint scheme, though, but its chameleon qualities might just persuade the neighbours that you change your car a lot. Note the driving lights set in the air intake in the lower valance.

The Alfa shield is elongated once more to previous proportions, and the slots that surround it in the front valance are reminiscent of the 1930s Monza. There's a further subtlety about the 156 bonnet in the way the top of the shield rises slightly to blend into a tiny ridge that merges with the bonnet.

The Spider and GTV models were introduced in the UK in 1995 and shared the same running gear (inherited from the later 155) and body panels as far as the windscreen, and their distinctive twin headlights were the product of a single unit hidden behind the bonnet. The key stylistic feature was a steep groove that extended steeply upwards along the rounded flanks to the high sawn-off tail of the GTV. The luxuriously appointed two-plus-two cabin featured a large, steeply raked rear window incorporating a third stop light at its base.

The 16-valve 2-litre GTV TwinSpark pictured here preceded the 24-valve 3-litre six-speed model by a couple of years. These cars had fabulous styling that was quite avant-garde when they were launched. They were fast and exhilarating to drive on smooth blacktop roads, but marginally less so on inferior surfaces. That said, they displayed none of the scuttle shake that I experienced when researching a book on their Spider sister, which hated poor road surfaces. Inside the GTV cabin, creature comforts included an automatic climate control system with pollen filter, while upholstery and trim could be specified in blue, white or red Momo leather which induced feelings of longing – for that elusive lottery win.

Just as the 164 proved that Alfa Romeo could make an executive model without tears, so the 166 consolidated that view and made the case even stronger. It was a late starter, however, as Fiat boss Paolo Cantarella vetoed the original designs in order to give the 164's replacement the optimum chance of cracking the German-dominated executive sector. When it was launched in 1998 changes to the suspension and interior lifted it head and shoulders above the dreary Aryan tin-tops.

The ergonomics of the 166 are simply exquisite. At the time of writing, the model had only just been launched and I sat in one on the Alfa stand at the 1998 Birmingham Motor Show. Its leather seats are comfortable and fully electrically adjustable, while the instrumentation includes computerised satellite navigation aids, cruise control, a 'smart' radio that adjusts volume automatically, and speedo and rev-counter overlaid in the semi-circular instrument binnacle. There's a passenger-side airbag too, and as you'd expect, the interior exudes an air of quality.

The family resemblance with the smaller 156 is clear, but the 166 has colour-coded bumpers and the rubbing strips are absent, giving it a purer appearance. There's still an ample amount of detailing to take in, though. The rear lights echo the narrow strip clusters of the 156, but the sculpted indent that runs along the sides is given more prominence in the 166. 'Telephone-dial' alloy wheels are fitted on the base-model 2-litre TwinSpark, but different options are likely for other models in the range – the 2.5-litre V6 and 3-litre V6. The four-cylinder car just gets the five-speed manual box, while the 2.5 has the option of four-speed ZF automatic with the Sportronic (a sort of halfway house) shift. The 3-litre car comes with either ZF auto or six-speed manual gearbox. There's also a 'sport throttle response', which modifies the induction system's butterfly valve opening and is meant to enhance the sporting experience. Whether it does this or not, I couldn't say.

There's always some corporate crossover to ease the financial burden of car manufacture, whether it's an engine and gearbox package or communal suspension system, and in the case of the 166, it shares its floorpan with the Lancia Kappa. Alfa's limo has a suspension set-up very similar to that of the 156, but the multi-link rear end has been developed using a mixture of stainless steel, cast-iron and aluminium components. The 166's comprehensive specification includes anti-slip and traction control, and ABS brakes with active sensors.

Mindful of its sporting pedigree and recent touring car success with the 155, Alfa Romeo produced two competition versions of the 156 from the outset. One was the highly developed car that took part in the 1998 Italian Superturismo series, and the other was the more standard Group N car, which uses the 2-litre TwinSpark motor. Much of the preparation for the Group N versions was carried out at the factory, including stripping out the interiors and fitting roll-over cage, racing seat and six-point safety harness, fire extinguisher, competition springs and dampers, strut brace, ignition cut-out and bonnet fasteners – all for the price of the regular road car. Also included were the sections of aerodynamic body kit like the rear wing and air dam.

The regulations under which the 156 Superturismo races permit far more extensive alterations to the standard car – like the UK's BTCC where the 155 scored in 1994. The 156 Superturismo power unit is based on the crankcase and cylinder head of the regular engine, lubricated by a dry-sump system and using specially made manifolds, pistons, cams and cranks, and it develops 300bhp at 8200rpm. Like the 155, the current engine's internal dimensions have been altered to give a square 86mm × 86mm configuration. A specially made engine management system controls electronic ignition and injection system, and the whole unit is set lower and further back in the engine bay to improve weight distribution. A six-speed X-Trac sequential shift is allied to a self-locking diff and viscous coupling, and the rose-jointed suspension set-up includes driver-adjustable anti-roll bars. The Superturismo runs on 19in magnesium alloy wheels and its progress is retarded by mighty 335mm Brembo ventilated discs with eight piston callipers. Drivers in the Italian series in 1998 were Nicola Larini and Fabrizio Giovanardi, and it is just a matter of time before UK race fans can see the 156 in action on British circuits.

Acknowledgements

My sincere thanks go to the following people who generously lent photographs for this book. First and foremost is Elvira Ruocco at the Alfa Romeo Centro Storico Documentazione at Arese, who provided much of the archive material, which makes this book a comprehensive look at Alfa's saloon car production. Thanks to the staff at Alfa GB for liaison, including Angi Voluti, Giulietta Calabrese and Puneet 'Josh' Joshi, who passed on the current press material. Also to Bob Hoare at Fiat UK, whose 2000 Berlina may still be for sale. Other pictures came from Alfasud specialist and Auto Italia Championship winner Ian Brookfield, who also has an impressive collection of Alfa Romeo sales brochures.

The Giulietta Berlina ti pictures came from John Brittan; Patrick Walter sent me photos of his 146 ti. My daughter Keri gave us a guided tour of Cumbria in a quest for locations to snap the family 155 V6. Richard Everton provided shots of his Giulia TI/Super and regular Super, which had received remedial treatment at the hand of the master, Mike Spenceley, who specialises in the restoration of classic Alfa Romeos at MGS Coachworks, Purley. Norwich Alfa specialist and former Alfasud protagonist Richard Drake of Richard Drake Motors and Alastair Kerridge of the Norwich Alfa dealership Lindfield Italia Ltd also supplied pictures. Michael Lindsay of the Alfa Romeo Owners' Club provided the Arna photograph, while Gary Chaplin sent pictures of his 156 V6 Sport 3 and Stuart Taylor forwarded shots taken at Assen, Holland.

Roger Monk, keeper of the 102- and 106-Series register, delved through his personal archives for pictures of the relevant Berlinas. Dave Benbow of Prodrive provided some insights into the 1995 BTCC season, as did Laurie Caddell of CW Editorial. Thanks also to Laurie for his GTA pictures, and to Pete Robain for the 1600 GT Junior pix. The 156 Superturismo photos are by Darren Heath. Other photographers whose work is featured are Steve Jones and Martin Ruddick.

Thanks also to Simon Fletcher, Annabel and Sarah at Sutton Publishing for providing the opportunity to give the Alfa saloons an airing in print! I'd like to dedicate the book to my daughter Zoë, who's becoming a real car buff.

INDEX

Abarth, 128, 130
Alfa Romeo cars (in chronological order):
P3, 12
6C 1750 Gran Turismo, 10, 11, 12
6C 2300, 12
8C 2300, 11, 12, 52
8C 2900, 13,
6C 2500/Villa d'Este, 6, 17, 18
158/159 Alfetta, 6, 13, 14, 89
6C 2500 Super Sport Freccia d'Oro, 21
6C 3000 CM/Disco Volante, 15, 22, 29
1900 Super Sprint, 21
1900 Berlina, TI and Normale, 6, 9, 15, 19, 20
1900 Sport Spider/2000 Sportiva, 21
Matta AR51 jeep, 23
BAT cars, 22, 115
Giulietta Berlina Normale/TI, 5, 6, 15, 23, 26, 27, 28, 29, 30, 31, 33, 39
Giulietta Sprint/Veloce, 6, 15, 22, 23, 24, 25
Giulietta 750-series Spider, 6, 15, 24
Giulietta 101-series Spider, 15, 27
2000 Spider 'Touring', 33
2000 Berlina, 6, 33
2000 Sprint 'Bertone', 33
2000 Sprint 'Touring', 38
Tipo 103: 32
2600 Spider 'Touring', 34, 35, 36
2600 Sprint 'Bertone', 34, 35, 36
2600 Sprint 'Zagato', 37
2600 Berlina, 6, 34, 35, 36
Sprint Speciale (SS), 15, 38
Giulietta SZ/SVZ, 15, 32, 142
Giulia 101-series Sprint, 23
Giulia 105-series 1300 TI, 6, 16, 47, 48, 49, 50
Giulia 105-series 1600 Super, 5, 6, 7, 16, 45, 46, 47, 48, 49, 51, 55, 56, 57, 80, 142
Giulia 105-series TI and TI/Super, 7, 15, 32, 39, 40, 41, 42, 43, 44, 45, 51
Giulia Sprint GT/GTV and GTA/GT SA/GT Am/GTA Junior, 5, 6, 7, 15, 39, 40, 44, 52, 53, 54, 55, 56, 61, 62, 63, 64, 73, 109, 110, 112, 139
Giulia GT-Z/TZ2, 15, 51, 52
1600 Duetto, 6, 7, 15, 40
Montreal, 60, 104
1750 Berlina, 6, 16, 55, 56, 57, 58, 59, 144
Junior Z, 61
2000 Berlina, 6, 16, 50, 57, 58, 59, 60
Tipo 33: 5, 15, 60, 62, 77, 104
Tipo 177: 14
Tipo 179: 14
Alfasud and variants, 6, 16, 65, 66, 67, 68, 69, 70, 71, 72, 73, 74, 75, 76, 77, 82, 83, 86, 87, 100, 105, 136
Giardinetta, 68, 79
Arna, 16, 81, 82, 141
Alfetta saloon, 6, 7, 16, 89, 90, 91, 92, 93, 94, 98, 99, 100, 103, 111, 126
Giulietta, 7, 16, 89, 94, 95, 96, 97, 100
Alfa 6: 6, 16, 98, 99, 110
Alfa 90: 7, 99, 100, 101, 110
Alfetta GTV/GTV6: 16, 77, 97, 104, 105, 110, 115
Alfa 33: 6, 16, 77, 78, 80, 81, 82, 83, 84, 85, 85, 86, 87, 88, 100, 137
Giardinetta/Sportwagon, 16, 79, 84, 88
Tipo 183T, 106
Alfa 75: 7, 8, 78, 100, 107, 108, 109, 110, 111, 112, 113, 114, 115, 116, 117, 118, 123, 125, 126, 144
Alfa 164: 8, 117, 118, 119, 120, 121, 122, 123
SZ/ES30, 115, 116
Alfa 155: 8, 123, 124, 125, 126, 127, 128, 129, 130, 131, 132, 133, 134, 135, 137, 144, 152, 156
Spider, 126, 152
GTV, 126, 152, 153
Alfa 145: 8, 136, 137, 138, 139, 140, 141
Alfa 146: 8, 137, 138, 139, 140, 141
Alfa 156: 8, 16, 117, 142, 143, 144, 145, 146, 147, 148, 149, 150, 151, 155, 156, 157
Alfa 166: 117, 154, 155
Alfa Romeo factories:
Arese, 6, 15, 21, 32, 39, 55, 90, 103, 104, 111, 127
Balocco test track, 15
Pomigliano d'Arco, 6, 79, 87, 88
Portello, 6, 9, 14, 15, 26, 27, 39
Pratola Serra, 81, 141
Alfa Romeo Owners' Club:
AROC Championship, 75, 76, 86, 103, 105, 113, 115
National Alfa Day, 7, 51, 52, 58, 90, 113
Alfa Romeo personnel:
Anderloni, Carlo Bianchi, 98
Chiti, Carlo, 15
Columbo, Gioacchino, 13
Ferrari, Enzo, 10, 11
Jano, Vittorio, 10, 11, 13
Limone, Sergio, 128
Masoni, Edo, 104
Merosi, Giuseppe, 9, 11
Pianta, Giorgio, 112, 128
Piccione, Alessandro, 126
Romeo, Nicola, 10, 14
Russo, Nini, 128
Satta, Orazio, 19
Stella, Ugo, 9
Andretti, Mario, 14
Ascari, Antonio, 10
Assen circuit, 43
Autodelta Racing Team, 15, 52, 54, 61, 62, 77, 110

Baldi, Mauro, 106
Bell and Colvill, 97
Bertone, Nuccio/Carrozzeria, 22, 24, 33, 38
Birmingham Motor Show (NEC), 154

BMW, 5, 44, 63, 117, 133
Boano, Mario/Carrozzeria, 22, 122
Bosch:
   ABS, 126
   distributors, 109
   fuel injection, 99, 101, 109, 126, 138
Brakes:
   Brembo, 129, 157
Brands Hatch circuit, 77, 86, 106, 115
British Touring Car Championship (BTCC), 128, 130, 131, 132, 133, 157
Brittan, John, 30
Brookfield, Ian, 2, 75, 76
Bussinello, Roberto, 61

Cadwell Park circuit, 76, 77
Carburettors:
   Dell'Orto, 82, 90
   Solex, 28, 31, 34, 69
   Weber, 24, 43, 51, 53, 64, 69, 74, 82
Carrera Messicana, 19
Castagna, Ercole/Carrozzeria, 13
Castle Combe circuit, 76
Cheever, Eddie, 14
Colli, Carrozzeria, 15, 29, 79
Coys of Kensington, 44

Dallara F3, 109
Darracq, 9
De Adamich, Andrea, 42, 44, 54, 60, 77
De Cesaris, Andrea, 14, 106
Depailler, Patrick, 14
Di Bono, Riccardo, 44
Dini, Spartaco, 54
Donington Park circuit, 135

European Touring Car Championship, 44, 61, 62
Everton, Richard, 41, 46

Fangio, Juan-Manuel, 14, 29, 89
Farina, Nino, 14, 89
Farina, Stabilimenti, 13, 22, 34
Ferrari, 10, 11, 12, 13
Fiat, 9, 14, 16, 118, 122, 130, 131, 142
   Cantarella, Paulo, 154
   Centro Stile, 22, 122

Da Silva, Walter, 122
Fiat models:
   124 Coupé, 122
   Croma, 8, 118
   Tempra, 124
Frankfurt Motor Show, 30, 117

Galli, Nanni, 54, 61, 62, 77
Garrett turbocharger, 97, 112
Ghia, Carrozzeria, 18, 22
Giovanardi, Fabrizio, 157
Giugiaro, Giorgetto, 36, 104
Giunti, Ignazio, 54
Goodwood circuit,

Harvey-Bailey, Rhoddy, 110
Heels, Graham, 86
Hezemans, Toine, 63, 77
Hruska, Dr Rudolf, 87

International Touring Car Championship (ITC), 121, 134, 135

Jaguar, 5, 44

Kamm, Prof. Wunibald (Kamm-tail), 7, 34, 52, 137
Kay, Roger, 115
Kyalami circuit, 34, 106

Lancia, 8, 9, 13, 14, 16, 22, 88, 112, 118, 124, 128, 129, 130, 155
Larini, Nicola, 113, 135, 157
Le Mans 24-Hours, 11
Long Beach GP, 106
Lotus Cortina, 5, 7, 41, 54
Lucas distributors,

Mallory Park circuit, 105
Mercedes-Benz, 134
Mille Miglia, 13, 15, 23
Mini Cooper, 44, 46
Monza, 11, 42, 44, 45
Moss, Stirling, 60
Munaron, Gino, 44

Nürburgring, 12,

Opel, 134
Osi, Carrozzeria, 35

Patrese, Riccardo, 14
Perkins diesel engine, 51

Picchi, Gian-Luigi, 62, 63
Pininfarina, Giovan-Battista/ Sergio/Carrozzeria, 16, 24, 79
Porsche, 77, 87
Presley, Graham, 113
Prodrive, 132

Renault, 32
Restoring Classic Cars magazine, 50
Rindt, Jochen, 54
Russo, 'Geki', 52

Saab, 8, 118
Sebring 12-Hours, 54
Silverstone circuit, 41, 44, 62, 63, 131
Simoni, Giampiero, 128, 131, 132, 133
Snetterton circuit, 54, 61, 75, 110
Spa Francorchamps circuit, 44
Spenceley, Mike, 41
Spica fuel injection, 62, 90
Squadra Blanca, 43
Stewart, Jackie, 54, 62
Streather, Dave, 82

Targa Florio, 9, 10, 19, 52
Tarquini, Gabriele, 113, 128, 131, 132, 133, 135
Thruxton circuit, 133
Touring-of-Milan, Carrozzeria/Superleggera, 13, 17, 21, 22, 33, 34, 38, 98
Turin Motor Show, 37, 66
Tyres:
   Michelin, 129
   Pirelli Cinturato 29
   Pirelli P-Zero, 116
   Yokohama, 114

Villamil, Luis, 133

Warwick, Derek, 132
Wheels:
   Campagnolo, 95
   Speedline, 123, 129
   OZ, 134

Zagato, Elio/Carrozzeria, 13, 32, 37, 51, 61, 79, 115, 142
Zandvoort circuit, 63, 106
Zeccoli, Teodoro, 52